U.S. Department of
ENERGY

I0468986

Uranium Enrichment Decontamination and Decommissioning Report

Report to Congress
December 2010

United States Department of Energy
Washington, DC 20585

Message from the Assistant Secretary for Environmental Management

In accordance with section 1805 of the Atomic Energy Act of 1954, as amended by the Energy Policy Act of 1992, the Department of Energy is providing the sixth Triennial Report to Congress on the progress of the Uranium Enrichment Decontamination and Decommissioning Fund (UED&D Fund). In the last report submitted to Congress in January 2008, the Department concluded that in no case was the UED&D Fund sufficient to complete the cleanup activities at the three Gaseous Diffusion Plants located in Oak Ridge, Tennessee; Piketon, Ohio; and Paducah, Kentucky. The January 2008 report estimated that the UED&D Fund shortfall was about $10.9 billion, and the fund would be exhausted by 2022.

This report updates the Congress on the progress and successes of the cleanup over the last three years. The UED&D Fund remains insufficient. The shortfall is now estimated at $11.8 billion, with the fund predicted to be exhausted by 2020. However, there are inherent uncertainties associated with planning for large complex projects and as they progress through their life cycle, the Department's goal is to reduce this uncertainty with higher quality work estimates, good faith regulatory negotiations and project management oversight and controls. To satisfy the requirement that DOE submit triennial reports to Congress, this report is being provided to the following Members of Congress:

- **The Honorable Daniel K. Inouye**
  Chairman, Senate Appropriations Committee

- **The Honorable Thad Cochran**
  Ranking Member, Senate Appropriations Committee

- **The Honorable Jeff Bingaman**
  Chairman, Senate Committee on Energy and Natural Resources

- **The Honorable Lisa Murkowski**
  Ranking Member, Senate Committee on Energy and Natural Resources

- **The Honorable Byron Dorgan**
  Chairman, Subcommittee on Energy and Water Development, Senate Committee on Appropriations

- **The Honorable Robert Bennett**
  Ranking Member, Subcommittee on Energy and Water Development, Senate Committee on Appropriations

- **The Honorable David Obey**
  Chairman, House Appropriations Committee

- **The Honorable Jerry Lewis**
  Ranking Member, House Appropriations Committee

- **The Honorable Henry A. Waxman**
  Chairman, House Committee on Energy and Commerce

- **The Honorable Joe Barton**
  Ranking Member, House Committee on Energy and Commerce

- **The Honorable Bart Gordon**
  Chairman, House Committee on Science and Technology

- **The Honorable Ralph M. Hall**
  Ranking Member, House Committee on Science and Technology

- **The Honorable Peter J. Visclosky**
  Chairman, Subcommittee on Energy and Water Development, House Committee on Appropriations

- **The Honorable Rodney Frelinghuysen**
  Ranking Member, Subcommittee on Energy and Water Development, House Committee on Appropriations

- **The Honorable Jim Bunning**
  U.S. Senate, State of Kentucky

- **The Honorable Mitch McConnell**
  U.S. Senate, State of Kentucky

- **The Honorable Ed Whitfield**
  U.S. House of Representatives, State of Kentucky, 1st District

- **The Honorable Sherrod Brown**
  U.S. Senate, State of Ohio

- **The Honorable George V. Voinovich**
  U.S. Senate, State of Ohio

- **The Honorable Jean Schmidt**
  U.S. House of Representatives, State of Ohio, 2nd District

- **The Honorable Michael Turner**
  U.S. House of Representatives, State of Ohio, 3rd District

- **The Honorable Lamar Alexander**
  U.S. Senate, State of Tennessee

- **The Honorable Bob Corker**
  U.S. Senate, State of Tennessee

- **The Honorable Zack Wamp**
  U.S. House of Representatives, State of Tennessee, 3rd District

- **The Honorable Lincoln Davis**
  U.S. House of Representatives, State of Tennessee, 4th District

If you have any further questions, please contact me or Mr. Steve Lerner, Office of Congressional and Intergovernmental Affairs, at (202) 586-5450.

Sincerely,

Inés R. Triay

Enclosure

Executive Summary

As required by the Energy Policy Act of 1992 (EPAct), the U.S. Department of Energy (the Department) is pleased to present to Congress the sixth triennial report providing analysis regarding sufficiency and management of the Uranium Enrichment Decontamination and Decommissioning (UED&D) Fund (the Fund). The Fund's primary mission is to provide decontamination and decommissioning (D&D) and cleanup of the nation's three gaseous diffusion plants (GDPs), namely the GDP at East Tennessee Technology Park (ETTP) in Oak Ridge, Tennessee; the Paducah GDP in Paducah, Kentucky; and the Portsmouth GDP near Piketon, Ohio.

The task of completing decontamination, decommissioning, and remedial action projects involves a large complex of interconnected facilities contaminated with industrial, chemical, nuclear, and radiological hazardous materials. The primary beneficiaries are the public and workers who reside in the immediate areas surrounding each site. The Department looks forward to sustaining and completing this vital program.

Since the establishment of the Fund in 1992, the Department has cleaned out three of the 12 massive process buildings (K-29, K-31, and K-33 at ETTP) and demolished the K-29 Building. D&D has been completed on K-31, but planning for demolition has not been completed. K-33 will be demolished using funding provided under the American Recovery and Reinvestment Act (ARRA). Another significant milestone was completed in early 2010 when demolition was completed on the half-mile long west wing of the U-shaped K-25 Building at ETTP. The west wing of the K-25 Building had a footprint of approximately 20 acres. The process buildings at Paducah are still leased by the United States Enrichment Corporation (USEC) and are not yet available for D&D. Therefore, the Department has focused on other cleanup work that can be conducted without interfering with USEC's operations under the lease. At Portsmouth, USEC returned certain leased support facilities to the Department, and the Department began D&D activities. Across the three GDP sites, the Department has completed cleanup and D&D of hundreds of excess support facilities, undertaken numerous remedial actions involving soil and groundwater, and disposed of millions of cubic feet in waste materials.

The American Recovery and Reinvestment Act of 2009 (ARRA) provided $390M of funding across the three sites to accelerate this important program. The Department is removing inactive facilities, remediating soils, and providing reimbursement of remediation costs for uranium and thorium processing sites. Also, the sale of excess uranium provided $100M in fiscal year (FY) 2010 to help accelerate specific preparatory work at Portsmouth.

Although cleanup and D&D progress has been significant at all three sites, much work remains. The estimate to complete ETTP is approximately $2.1 billion with work expected to be completed in FY 2020. Portsmouth began D&D on a limited scale in fiscal year FY 2009 with ARRA funding, with the initial Department of Energy D&D contract for Portsmouth awarded in August 2010. The Department received the Portsmouth process buildings from USEC on

September 30, 2010. Transfer of formerly leased facilities from USEC to the Department will continue for the next few years as hundreds of facilities are returned to the Department. Under the current conceptual design, cleanup is expected to be completed by 2044. The cost estimate for the Portsmouth D&D project is conceptual and has a most probable value of $7.5 billion in year of expenditure dollars. The Paducah process buildings are still leased and operated by USEC and are not projected to be available for D&D until 2017. Cleanup is expected to be completed by 2040. The conceptual cost estimate for the Paducah D&D project has a most probable value of $9.0 billion in year of expenditure dollars. A conceptual estimate for cost and schedule contains a high level of uncertainty and is expected to be within +50/-30% of the actual costs, given the assumptions made in the estimate. The Fund must also continue reimbursements to licensees of active uranium and thorium processing sites for the portion of their remedial action costs attributable to Federally related byproduct material. The cost estimate for uranium and thorium reimbursements is $291 million.

Section VIII "Fund Analysis" of the report examines the Fund's sufficiency. The Department has established a "Base Case" that reflects the most likely scenario for completing the cleanup mission. It reflects current programmatic assumptions regarding scope, schedule, cost, etc. Given the $4.9 billion current Fund balance and accounting for future inflows to the Fund from interest earnings, the Department concludes that the Fund will have a shortfall of $11.8 billion to complete the GDP cleanup activities. Without additional deposits into the Fund the current balance in the Fund is projected to be exhausted in 2020.

This projection is comparable to the analysis from the Initial Fund Assessment completed by the General Accounting Office (GAO) in November 1991, before the Fund was established. GAO analyzed multiple scenarios to assess the adequacy of a $500 million annual deposit into the Fund to cover cleanup costs at the three GDPs. GAO estimated that for the Fund to be sufficient to cover all cleanup work, it would need to receive annual deposits of $500 million indexed for inflation, for the life of the cleanup work (approximately 2040). However, the EPAct set an annual contribution level of $480 million indexed for inflation for 15 years. The Department was to formally assess the Fund's sufficiency at the end of the 15 years of contributions and determine if the Fund should be reauthorized, which it did in the fifth Triennial Report dated January 2008. The amount to be collected over these 15 years would total $7.2 billion in 1992 dollars, which is significantly less than the $19.1 billion cost estimate that existed at that time.

Uranium Enrichment Decontamination and Decommissioning Report

Table of Contents

Acronyms

ACP	American Centrifuge Plant
ARRA	American Recovery and Reinvestment Act of 2009
CAB	Citizens Advisory Board
CERCLA	Comprehensive Environmental Response, Compensation, and Liability Act of 1980
CSB	cold standby
CSD	cold shutdown
CSOU	Comprehensive Site-wide Operable Unit
the Department	U.S. Department of Energy
D&D	decontamination and decommissioning
DNAPL	dense non-aqueous-phased liquid
DQO	data quality objective
DU	Deferred Units
EM	Office of Environmental Management
EMWMF	Environmental Management Waste Management Facility
EPAct	Energy Policy Act of 1992
ETTP	East Tennessee Technology Park
FFA	Federal Facility Agreement
the Fund	Uranium Enrichment Decontamination and Decommissioning Fund
FY	fiscal year
GAO	Government Accountability Office (pre-2004 General Accounting Office)
GCEP	Gas Centrifuge Enrichment Plant
GDP	gaseous diffusion plant
KOW	Kentucky Ordnance Works
LTS	long-term stewardship
NAS	National Academy of Science
NPL	National Priorities List
OSWDF	on-site waste disposal facility
OU	operable unit
PCB	polychlorinated biphenyl
RCRA	Resource Conservation and Recovery Act of 1976
ROD	record of decision
S&M	surveillance and maintenance
SWMU	solid waste management unit
TSCA	Toxic Substances Control Act of 1976
USACE	U.S. Army Corps of Engineers
USEC	United States Enrichment Corporation
USEPA	U.S. Environmental Protection Agency

I.   Legislative Language

This report responds to legislative language set forth in the Atomic Energy Act of 1954, as amended by the Energy Policy Act of 1992, in Section 1805, wherein it is stated:

> Within 3 years after the date of the enactment of this title, and at least once every 3 years thereafter, the Secretary shall report to the Congress on progress under this chapter. The 5th report submitted under this section shall contain recommendations of the Secretary for the reauthorization of the program and Fund under this title.

II.   Introduction

Gaseous diffusion is one of several uranium isotope separation technologies that were developed as part of the 1940s Manhattan Project.   The GDPs were constructed by the United States of America, the Soviet Union, the United Kingdom, France, and China. Most of the GDPs have long since shut down, unable to economically compete with newer enrichment techniques. Three GDP sites exist in the United States: 1) East Tennessee Technology Park (ETTP) in Oak Ridge, Tennessee; 2) Paducah in Paducah, Kentucky; and 3) Portsmouth in Piketon, Ohio. Of these three GDP sites, only Paducah is operating today.

In 1992, the U.S. Congress law passed the Energy Policy Act of 1992, as amended (EPAct), which created the United States Enrichment Corporation (USEC), a Government corporation with the mission of restructuring the Government's uranium enrichment operations that has since been privatized. The EPAct also established the Uranium Enrichment Decontamination and Decommissioning (D&D) Fund (the Fund) to provide the necessary resources to clean up the environmental liability created through operations of gaseous diffusion facilities.

The gaseous diffusion process requires the enrichment of uranium in a gaseous state, which periodically leads to the release of contaminated gases. The preparation of uranium hexafluoride (hex) feedstock was the first application for commercially produced fluorine. Significant problems were encountered and technical solutions were subsequently developed in handling both fluorine and the corrosive hex gas.   Also, during the course of operating and maintaining the GDPs, a large infrastructure of support facilities such as storage buildings, cleaning shops, and laboratories were required to handle other radioactive or hazardous materials. Lastly, since the earliest days of the Manhattan Project, an accepted waste management practice at these plants was the on-site disposal of waste generated by the GDPs in unlined trenches or bore holes. While protective of human health and the environment under standards in place at that time, far more stringent environmental standards exist today.

The materials handled at the GDPs, the "learning curve" for handling them, and the waste management practices of the time contributed to the contamination of environmental media (i.e., soil, sediments, and groundwater) at these sites. This contamination is the focus of past, present, and future remedial actions that are paid by the Fund.

The U.S. Department of Energy (the Department) is the lead agency managing D&D activities at all three sites under the Comprehensive Environmental Response, Compensation, and Liability Act of 1980 (CERCLA). The Oak Ridge Reservation and Paducah GDP are listed on the U.S. Environmental Protection Agency's (USEPA's) National Priorities List (NPL) and have negotiated Federal Facility Agreements with their State and Federal regulators. Environmental Remediation at Portsmouth is regulated by the Ohio Environmental Protection Agency under the Resource Conservation and Recovery Act of 1976 (RCRA), and has negotiated a Consent Order and Consent Decree and a Directors' Final Findings and Orders agreement with the State. Portsmouth will conduct D&D activities under the CERCLA Removal Action process Remedial Investigation/Feasibility Study (RI/FS) through Remedial Design/Remedial Action (RD/RA) process for other structures.

This introduction provides general background information on the scope of activities that the Fund finances. Subsequent sections provide the history, regulatory basis, cleanup plan summary, and challenges and uncertainties for each site. The remaining section addresses the status of the Fund's current resources and likely future resource needs. The report concludes with recommendations, followed by appendices with details of the financial analysis.

Background

In 1992, the United States Congress enacted the EPAct, creating USEC and establishing the Fund. Though privatization of the enrichment enterprise was an important feature of the EPAct, one of the most challenging aspects of the law was its mandate to address the cleanup liability of past enrichment operations at Department facilities. The cleanup of these facilities remains the responsibility of the Department. In an effort to address the liability issue, the Fund was established to provide for ultimate D&D of the three GDPs; remedial actions at the sites to the extent the Fund is sufficient; management of waste generated by historical operations; uranium/thorium licensee reimbursements; and eventual disposition of the depleted uranium hexafluoride (DUF_6) cylinders. [Note: In fiscal year (FY) 2001, Congress provided funds from a separate appropriation and program for the disposition of DUF_6 cylinders.]

The relevant portions of the EPAct are shown in Figure 1, including the Department's fiscal and managerial responsibilities for cleanup activities.

Fund and Program Requirements Summary
Energy Policy Act of 1992

Section 1801(a)

ESTABLISHMENT.-There is established in the Treasury of the United States an account to be known as the Uranium Enrichment Decontamination and Decommissioning Fund (referred to in this chapter as the 'Fund'). The Fund, and any amounts deposited in it, including any interest earned thereon, shall be available to the Secretary subject to appropriations for the exclusive purpose of carrying out this chapter.

Section 1802

(a) AMOUNT.-The Fund shall consist of deposits in the amount of $480,000,000 per fiscal year (to be annually adjusted for inflation using the Consumer Price Index for all-urban consumers published by the Department of Labor) as provided in this section.

(b) SOURCE.-Deposits described in subsection (a) shall be from the following sources:
(1) Sums collected pursuant to subsection (c)... (2) Appropriations made pursuant to subsection (d).

(c) SPECIAL ASSESSMENT.-The Secretary shall collect a special assessment from domestic utilities.

(d) AUTHORIZATION OF APPROPRIATIONS.-There are authorized to be appropriated to the Fund, for the period encompassing 15 years after the date of the enactment of this title, such sums as are necessary to ensure that the amount required under subsection (a) is deposited for each fiscal year.

Section 1803

(b) PAYMENT OF DECONTAMINATION AND DECOMMISSIONING COSTS.-The costs of all decontamination and decommissioning activities of the Department shall be paid from the Fund until such time as the Secretary certifies and the Congress concurs, by law, that such activities are complete.

(c) PAYMENT OF REMEDIAL ACTION COSTS.-The annual cost of remedial action at the Department's gaseous diffusion facilities shall be paid from the Fund to the extent the amount available in the Fund is sufficient. To the extent the amount in the Fund is insufficient, the Department shall be responsible for the cost of remedial action.

Section 1805

Within 3 years after the date of the enactment of this title, and at least once every 3 years thereafter, the Secretary shall report to the Congress on progress under this chapter. The 5th report submitted under this section shall contain recommendations of the Secretary for the reauthorization of the program and Fund under this title.

Figure 1. Relevant portions of the EPAct

Fund Revenues

The EPAct assigns the liability for past operations to the historical beneficiaries of the enrichment activities and provides direction for contributions that would be accumulated toward satisfying this liability. Historical beneficiaries of the enrichment process were United States utilities that purchased uranium from the Department's enrichment program and the Government's defense enrichment mission. Therefore, the Fund was designed to include annual contributions from utilities and contributions from the Government to cover the entire liability of GDP D&D and cleanup programs. Utility contributions were based on historical purchases of enrichment services, which were measured in Separative Work Units (SWU). The utility contributions accounted for approximately one-third of the annual contributions, while Government contributions accounted for the remaining two-thirds.

Originally, the EPAct authorized annual deposits to the Fund of $330 million in Government contributions and $150 million in domestic nuclear utility contributions. These contributions were to be made for 15 years beginning in FY 1993 and adjusted annually for inflation. Total collections were to equal $2.25 billion from the utilities and $4.95 billion from the Department, also adjusted annually for inflation.

Fund balances in excess of current-year funding needs are to be managed and invested to earn interest from Government securities. The design of the Fund also includes fiscal oversight and other accounting measures. The EPAct requires progress reports to Congress every three years and an annual audit by the Department's Office of the Inspector General. The Department manages the Fund as it was designed by Congress.

Fund Expenditures

The Fund includes provisions for the D&D of the three GDPs and for remedial action cleanup to the extent the Fund is sufficient. The Fund also provides that owners of uranium and thorium milling sites that provided ore to the government were to be reimbursed for a portion of the cleanup expenses at their sites.

Since the establishment of the Fund in 1992, the Department has cleaned out three of the 12 massive process buildings (K-29, K-31, and K-33 at ETTP) and demolished the K-29 Building. D&D has been completed on K-31 and K-33. Demolition of K-31 has not yet been planned; however, K-33 is to be demolished using ARRA funding. Another significant milestone was completed in early 2010 when demolition was completed on the half-mile long west wing of the U-shaped K-25 Building at ETTP. The west wing had a footprint of approximately 20 acres. The process buildings at Paducah are still leased by USEC and are not yet available for D&D. Therefore, the Department has focused on other cleanup work that can be conducted without interfering with USEC's operations under the lease. At Portsmouth, the Department retook possession of certain leased facilities from USEC and has undertaken D&D activities. Across the three GDP sites, the Department has completed cleanup and D&D of hundreds of excess support facilities, undertaken numerous remedial actions involving soil and groundwater, and disposed of millions of cubic feet in waste materials.

In addition, USEC has ceased uranium enrichment at the Portsmouth GDP and the Department has been contracting with USEC to perform prep work in the leased facilities that will aid future D&D by the Department. For example, USEC has been removing uranium deposits from the process lines, which will significantly reduce safety risks and costs during future D&D work.

The ETTP D&D is scheduled to be completed by 2020. The Portsmouth D&D was initiated in 2009 with American Recovery and Reinvestment Act (ARRA) funding, and will generate increased expenditures from the Fund with the transition to the D&D contractor. Although the Paducah site is still enriching uranium, the Department continues to plan for D&D when operations are concluded. Substantial remedial action and waste disposition activities have been conducted at all three GDPs. The Department has been making uranium/thorium reimbursements throughout the life of the Fund with $605.3 million paid through FY 2009

($573.4 from the Fund and $31.9 from ARRA). The Department projects that future payments under this reimbursement program will continue through FY 2024.

The Department, GAO, and other outside entities have provided numerous studies and cost estimates that stressed the need for continued funding analysis. The Department has implemented many recommendations of these previous studies and cost estimate reviews, and has made significant changes with the benefit of the experience gained from conducting D&D projects at the GDPs and across the Department's Environmental Management (EM) Program. More importantly, independent cost estimates for the Portsmouth and Paducah sites were developed in FY 2006 to address comprehensively the entire D&D and cleanup liability at those sites.

Impact of Previous Reports and Estimates

In September 1991, prior to the passage of the EPAct, the Department estimated it would cost $21 billion (FY 1992 dollars) and take more than 40 years to accomplish cleanup of the GDPs. The estimate was $19.1 billion with depleted uranium hexafluoride (DUF_6) cylinder conversion excluded as it is today. Those preliminary estimates, mostly developed using parametric models, assumed cleanup to background levels and lacked the benefit of current characterization data. The amount of the preliminary estimates is comparable to the estimated costs reflected in the base case.

Using the original estimate, GAO estimated the cleanup would require $500 million per year, indexed to inflation, over the 40-year life of the cleanup. However, the EPAct authorized annual contributions of only $480 million (indexed to inflation) through 2007, for a total of $7.2 billion. Thus, when Congress established the Fund, there was a difference of $13.8 billion (FY 1992 dollars) between the estimated costs for D&D, remedial actions, and disposition of depleted uranium (a D&D activity under the EPAct), and the anticipated contributions to the Fund. Eliminating the cost for DUF_6 conversion (now funded separately), there remained a difference of $11.9 billion between projected costs and funds authorized by the EPAct.

As required by the EPAct, the National Academy of Science (NAS) conducted a study of the program and published its recommendations in their 1995 report *Affordable Cleanup? Opportunities for Cost Reduction in the Decontamination and Decommissioning of the Nation's Uranium Enrichment Facilities*. The purpose of the NAS report was to identify major cost reduction opportunities for the project. The GAO also issued a report in FY 2004 that investigated the adequacy of the Fund to cover authorized activities at the three GDPs. In its report, *Uranium Enrichment Decontamination and Decommissioning Fund is Insufficient to Cover Cleanup Costs* (GAO-04-692), GAO concluded that, despite the Department's efforts to reduce costs (including recommendations from the NAS study), and based on GAO's assumptions and projections of costs and revenues, the Fund would not be sufficient to cover the expected cleanup costs. Further, GAO estimated a shortfall of revenue between $3.5 and $5.7 billion (in 2004 dollars). GAO recommended Congress consider reauthorizing the Fund for an additional three years and that the Department reassess the Fund's sufficiency during that period to determine if additional extensions were necessary. Also, GAO recommended that the Department develop life-cycle cost and schedule plans to accomplish D&D at the Portsmouth and Paducah sites.

The Department issued triennial reports to Congress beginning in 1995. The 2007 report was the fifth triennial report and as required by the EPAct, included a formal analysis of the Fund's sufficiency. The 2007 report projected the Fund to be insufficient by approximately $10.9 billion.

In an effort to maximize the efficiency of the entire GDP D&D Program, the Department has implemented key recommendations from all of these past reports, including those from NAS and GAO. For example, in 2006, the Department completed a comprehensive plan, schedule, and cost estimate for D&D at Portsmouth. Other implemented recommendations include stakeholder and regulator involvement, as well as, specific plans for acquisition of a D&D contractor through an open competitive procurement process. These elements, as well as, updated strategies for cost and risk reduction, are included in this report. The Department has also taken the opportunity to benefit from lessons learned through past and ongoing D&D projects across the Department complex (e.g., ETTP, Fernald Environmental Management Project, Mound, Rocky Flats, etc.).

III. Program Elements

The following sections detail program elements covered by the Fund.

Decontamination and Decommissioning

The Department's facilities are designated for D&D when they are no longer needed for existing missions. Buildings and facilities are scheduled for demolition based on either the most cost-effective schedule or the need to address risk-based safety concerns. Typical site-related contaminants within the GDP buildings include:

- Radioisotopes stemming from the historical enrichment process;
- Hazardous chemicals (e.g., TCE, PCBs, beryllium, etc.);
- Uranium;
- Technetium;
- Asbestos; and
- Other hazards typical of industrial facilities.

Liability for the D&D and remedial action beneath or peripheral to the building (as required to meet end-state goals) is incorporated into the Fund estimate presented in this Report. Also, the availability of valuable assets such as equipment or scrap metal is evaluated to determine if it can be used to offset the cost of facility decontamination on a case-by-case basis in accordance with the Secretary's recycling policy issued in 2001.

Remedial Action

Remedial actions involve assessment and cleanup of formerly used waste sites at the GDPs and subsurface media contaminated by historical GDP releases and/or operations. Burial grounds, disposal areas, holding ponds, pipeline leaks, and surface spill areas contribute to the contamination of soil, sediment, surface water, and groundwater with organic compounds and radionuclides. Remedial actions address the sources of contamination as well as the contamination in these environmental media. Priority is given to mitigating potential risks to site workers, off-site receptors, and environmentally sensitive areas.

The regulatory strategies at Portsmouth, ETTP and Paducah integrate the D&D with remedial actions. Required investigations and remediation of slabs and subsurface media following D&D of contaminated structures are closely linked to and are part of agreements with state and Federal regulators.

Waste Management

Waste management includes the management of waste generated from day-to-day operations, legacy waste previously generated at the GDPs and stored on site, and waste from current remedial actions at all sites. Cleanup generates additional waste that requires safe, efficient, and cost-effective disposition, including treatment and disposal. Waste management activities include treatment, storage, transportation, and disposal of transuranic and low-level radioactive waste, hazardous waste, mixed radioactive and hazardous waste, and sanitary waste in compliance with Federal, state, and local regulations and the Department's Orders.

Surveillance and Maintenance and Landlord Activities

The S&M activities encompass all actions required to ensure material, facility, and personnel safety and security. Facilities, equipment, and other systems S&M is required to mitigate the spread of contamination and protect human health and the environment. The GDPs' landlord activities support performance of the EM mission, are a necessary prerequisite to future response actions, and maintain the physical integrity of general-use facilities and infrastructure.

Uranium/Thorium Reimbursements

The EPAct Title X provides guidance to the Department for use of the Fund resources to reimburse licensees of active uranium and thorium processing sites for the portion of their remedial action costs attributable to federally-related byproduct material. Initial legislation directed a maximum reimbursement of $270 million for uranium licensees and $40 million for thorium licensees, plus adjustments for inflation.

Public Law 104-259, enacted October 9, 1996, increased maximum reimbursements to $350 million for uranium licensees and $65 million for thorium licensees, totaling $415 million. Public Law 105-388, enacted November 15, 1998, increased the maximum reimbursements for the thorium licensee to $140 million, bringing reimbursement to a total of $490 million. On August 21, 2002, Public Law 107-222 further increased the maximum thorium licensee reimbursements to $365 million, for a total of $715 million. The EPAct requires that annual payments be made to licensees. As of the FY 2009 payment, $605.3 million has been reimbursed ($573.4 from the Fund and $31.9 with ARRA funding).

IV. Strategy

The Department is committed to cleaning up ETTP and the Paducah and Portsmouth GDPs using resources provided by Congress through the Fund. As stewards of taxpayer dollars, the Department must perform cleanup in an efficient and cost-effective manner, with the specific goal of obtaining an end-state site condition acceptable to stakeholders. The Department intends to achieve this goal by establishing cleanup levels in partnership with regulators and site stakeholders that are commensurate with the future intended use of each GDP site, and using best-in-class contractors to execute the Department's cleanup mission. Performing this work is challenging and requires innovation and negotiation with open dialogue between all parties that are guided by the following primary tenets:

- Worker and public safety is the first responsibility and a zero-accident philosophy is the standard;
- Compliance with applicable laws and regulations is assured; and
- Support to National defense, security, science, or energy missions at each of the GDPs is enhanced by the cleanup mission, not hindered.

The Department's intent is to manage the GDP sites in an integrated manner. Each GDP has its own set of unique interfaces, and it is the Department's responsibility to assure all parties are engaged. While many of the key GDP interfaces are independent, the Department is still accountable to Congress and the other entities to complete the cleanup work safely and efficiently. Specific to this report, planning and executing work funded by the Fund requires open and direct dialogue with affected communities, state and Federal regulators, cleanup contractors, and other stakeholders, including the Congress.

The overall goal is to complete the GDP cleanup projects by 2044, which includes deactivation, waste management, S&M, D&D, and remedial actions. The Department is simultaneously working off the overall cleanup liability at all the GDPs in parallel; however, full-scale D&D of the GDP facilities will generally occur in the following sequence:

- ETTP D&D, started in 1994, will be complete in 2020.

- Portsmouth D&D, started in 2009, will be complete in 2044; and
- Paducah D&D will start in 2017 and be complete in 2040.

The Department has independent cost estimates to cover completion of all work scope covered by the Fund. The D&D and remedial action schedules for each site, as well as the overall Fund schedule, are tied to fundamental project management principles that include the following:

- Creating the vision of GDP cleanup;
- Developing the strategy necessary to achieve the vision;
- Prioritizing projects which assess high environmental or safety risk conditions;
- Sequencing activities consistent with sound engineering logic;
- Creating cost estimates and execution schedules based on bottom-up cost estimates and planning;
- Executing the scope fully consistent with the planning basis; and
- Using project management tools to monitor and adjust project performance as required.

Although a great deal of remediation and D&D has already been accomplished at the GDPs, significant challenges still exist. The Department is utilizing its risk management system to manage, mitigate, avoid, or eliminate the project risks (i.e., challenges) as well as control their potential impacts. The risk management system includes provisions for identifying, managing, and tracking risk elements. Key challenges include the following:

- In the past, the Department and its predecessors processed recycled fuel through the GDPs, which spread activation and fission products throughout the enrichment cascades. As a result, isotopes such as technetium pose unique challenges managing the D&D of facilities.
- When the Department shut down the Oak Ridge ETTP K-25 and K-27 buildings, uranium was not completely flushed from the system. As a result, retention of uranium is a risk in piping and process equipment that had to be addressed prior to general D&D. Lessons learned from ETTP D&D work are being shared with and addressed proactively at Portsmouth and Paducah.
- Considering the size of the GDPs, substantial characterization will be required to assure that efficient and compliant disposal is performed while the defined end-state goals of each site are met. The Department and its contractors will develop data quality objectives (DQOs) in partnership with the regulators to define and scope sampling campaigns necessary to compliantly and efficiently disposition waste.

Assumptions

The cleanup strategy for the GDPs is based on the following general assumptions:

- Uranium enrichment mission-related activities will be accommodated as a priority;
- An engineered, on-site radioactive, hazardous, and mixed waste disposal facility will be available for waste generated by D&D and remedial actions;
- Off-site Department of Energy and commercial disposal capacity will remain available;
- An adequate number of trained and skilled workers will be available to support the Fund completion schedule;
- Availability of security cleared workers will not be a hindrance to project execution;
- Sites will be cleaned to appropriately safe risk levels commensurate with end-state goals (limited land areas will require institutional controls following remediation);
- Initiation of D&D activities at Paducah and certain limited areas at Portsmouth are tied to the proposed schedule for implementation of centrifuge technology and may change based on production capabilities and market demand. D&D activities are also impacted at Portsmouth in terms of the schedule for the return of infrastructure facilities that support the centrifuge technology.
- Equipment and material removed from the buildings will be reused or recycled to the maximum extent practicable.

Most of these assumptions are already formalized and accepted. However, some will need to be reviewed and approved by the Department, the appropriate regulatory agencies, and other stakeholders.

Method of Accomplishment

All EM activities, with the exception of uranium/thorium reimbursements, are divided into manageable incremental projects through the use of work breakdown structures. Each incremental project has a well-defined end-point. Cost and schedule estimates are established and maintained through rigorous and formal change control procedures.

The Department will safely mitigate the liability posed by the GDPs in the following order of priority:

- Address and mitigate high risk buildings or remedial action projects;
- Remove legacy waste or materials stored in and around facilities;
- Deactivate facilities to assure the safety of future D&D;
- Remediate slabs and subsurface media as required;
- Remediate any other sources of radioactive or hazardous constituents;
- Remediate groundwater and any other contaminated subsurface media;
- Restore wetlands or perform any other appropriate resource conservation to meet end-state goals; and
- De-list facilities from USEPA's NPL.

The Department continues to improve a management approach that minimizes risk and maximizes cost savings and schedule control. The Department pursues efficient types of contracts and pricing mechanisms to allocate risk appropriately between the contractor and the Government. In addition, cost, schedule, and performance goals are controlled and monitored by an earned value management system. The current and future acquisition actions will accomplish the following:

- Demonstrate a risk analysis that minimizes technical complexity;
- Employ an acquisition strategy that appropriately and effectively uses competition, ties contract payments to accomplishments, and takes maximum advantage of commercial technology; and
- Monitor cost, schedule, and performance goals.

V.   East Tennessee Technology Park

History

ETTP is located on a 5,000-acre tract of land adjacent to the Clinch River, approximately 10 miles west of Oak Ridge, Tennessee. It was built as part of the World War II Manhattan Project to enrich uranium isotopes for the first atomic bombs. By the mid-1950s, five large uranium enrichment buildings covering 114 acres were in operation: K-25, K-27, K-29, K-31, and K-33. Four electrical switchyards and eight cooling towers served these buildings. Numerous support facilities were built where machinery was fabricated, serviced, repaired, and cleaned. Enrichment of weapons-grade uranium ceased in the 1960s. The plant enriched uranium for civilian nuclear power reactors until 1985, when all production operations ceased.

Uranium enrichment at ETTP has left a legacy of radioactive and chemical contamination in buildings, soils, sediments, and groundwater. The Department has identified more than 100 known or suspected sources of environmental contamination and has found uranium and other radioactive elements from enrichment processes to be widespread in the surrounding environment. Buried uranium-contaminated equipment and low-level radiological contaminated building rubble exist at several locations. Workers used volatile organic compounds in large quantities to clean and degrease equipment, resulting in the release of these compounds, specifically TCE, into the environment. These organic chemicals contaminated soil, surface water, and groundwater when they were spilled, burned in pits, discharged into holding ponds, or placed in trenches for disposal.

Regulatory Basis

In 1989, USEPA placed the Oak Ridge Reservation, including ETTP, on its NPL of contaminated sites. As a result, cleanup activities (some of which were initiated under RCRA) were to be completed as CERCLA remedial actions. In 1992, a Federal Facility Agreement (FFA) was executed among the Department, USEPA Region 4, and the State of Tennessee.

The FFA provides the framework for cleanup activities at ETTP, establishes enforceable milestones, and coordinates the requirements of CERCLA and RCRA. In 1995, the Department placed D&D activities under the FFA and CERCLA. Cooperative agency efforts and regulatory initiatives in conjunction with the involvement of community stakeholder groups (such as the Oak Ridge Site-Specific Advisory Board) help guide the process and assure cost-effective implementation of selected remedies.

Management of PCB waste is addressed through the Toxic Substances Control Act of 1976 (TSCA) Federal Facility Compliance Agreement between the Department and USEPA Region 4.

The Department in Oak Ridge is working with the State of Tennessee and USEPA Region 4 on a CERCLA decision strategy aimed at identifying remedial action objectives, cleanup criteria, land-use restrictions, and technologies to be used at ETTP. Initially, decisions were made on soils, subsurface infrastructure, and burial grounds. The ROD for these elements outside the fence, or Zone 1, was signed in November 2002. The ROD addressing these elements in Zone 2 was signed in April 2005, which includes the area within the main fence of ETTP (approximately 800 acres). Remedial activities required for the Zone 1 and Zone 2 RODs are currently underway. Residual contamination in groundwater, surface water, and sediments will be addressed on a site-wide basis. A Remedial Investigation/Feasibility Study and Proposed Plan are being finalized in support of the final ROD and closure of the site.

Cleanup Plan, Cost, and Schedule

The ETTP D&D Project includes demolition and disposal of the GDP process equipment, process buildings, and other facilities auxiliary to the gaseous diffusion process. The project also includes remediation of contaminated soils and groundwater associated with the gaseous diffusion operations.

The D&D work involves about 500 buildings and facilities covering approximately 15 million square feet, which are being addressed as CERCLA removal actions consistent with Departmental/USEPA policy guidance. To achieve the desired end-state for ETTP, buildings and facilities are scheduled for demolition based on the most cost-effective order. Facilities that have been officially transferred to a third party (currently the Community Reuse Organization of

East Tennessee) for commercial use before being demolished will be removed from the demolition schedule and EM program. The scope of other facilities' D&D includes planning, utilities deactivation, asbestos and hazardous material abatement, equipment disposal, structure demolition, and waste disposal.

Of the 5,000-acre ETTP footprint, there are 2,200 acres with the potential for unascertained amounts of contamination. There are known groundwater plumes from former burial grounds and contaminated soils. The cleanup strategy is to complete targeted remedial actions in Zone 1 (1,400 acres outside the fenced main plant area), including the groundwater plumes and potential contamination, facility decommissioning within the main plant area, and comprehensive remedial actions for the main plant subsurface area (i.e. Zone 2, which encompasses 800 acres inside the Main Plant area fence).

The estimate to complete ETTP is approximately $2.1 billion in year of expenditure dollars with approximately $2 billion from the Fund and $110 million from remaining ARRA funding ($118.2 million appropriated). Work should be completed by FY 2020. Appendix A includes a breakout of the required expenditures by fiscal year. The remaining major D&D work involves the four remaining process buildings (K-25, K-27, K-31, and K-33). The west wing of the K-25 building has been demolished, but D&D and demolition remain for the rest of the building. The K-27 building must still undergo D&D/demolition. ARRA funding is helping with preparation for K-27 D&D. ARRA funding will also be used to demolish the K-33 building. The Department is anticipating the K-31 building will be demolished and the cost estimates in this report account for demolition. In addition to the major process buildings, approximately 150 other support facilities still require D&D. Also, remedial action work remains, especially within the main plant area where D&D of facilities must occur before remedial action. There will be continuing post-closure, long-term stewardship activities that will be paid from funding sources other than the Fund.

Challenges and Uncertainties

The closure of ETTP has proven to be more complex and is taking longer to complete than previously planned. Technical challenges coupled with the changes in methods of accomplishment of the demolition of the K-25 and K-27 process buildings are the primary contributors to schedule delays and cost increases. The schedule for completing ETTP is now estimated to be FY 2020, which is eight years longer than the 2012 closure date assumed in the 2007 report. The extended schedule to complete the closure of ETTP adds infrastructure and support costs that will be incurred while the cleanup work is being completed.

In 1964, uranium enrichment operations in the K-25 Building ceased. Use of the lower two units in the East Wing as a "purge cascade" (a method of removing impurities) for Tc-99 ceased in 1977, and the building was shut down. Minimal surveillance and maintenance was performed after shutdown since no future mission need for the facility had been identified. The building's deteriorating 44 acre roof has led to significant, ongoing deterioration of the inside of the building. After a worker fell through a degraded floor in 2006, the K-25 D&D

project was significantly slowed until the Department could re-plan the project. Revised plans utilize greater reliance on heavy equipment to demolish the building, while minimizing labor inside the building for equipment removal. However, the Department has had to spend additional time and money to shore up the facility and assure safe working conditions for the labor driven work that still has to be conducted inside the building.

Although a great deal of work has been accomplished on the K-25 Building, the deterioration of the building continues to accelerate. Also, as the cleanup work progresses, the Department is gaining a better understanding of numerous difficulties and challenges that must be addressed. For example, there are issues with Tc-99 contamination in the K-25 East Wing. During characterization of the K-25 West Wing, sampling for Tc-99 confirmed that Tc-99 was at trace levels. In the K-25 East Wing, however, Tc-99 is recognized as a potential contaminant of concern that will require additional characterization and could impact both the demolition approach and the follow-on waste disposal options for certain sections of the East Wing. There is one area of the K-25 East Wing with higher Tc-99 levels that is being handled as a separate phase of the project. This area will require adjustments to the demolition approach to avoid spreading contamination. The key will be determining the sections within the Tc-99 areas of K-25 where the Tc-99 contamination level meets the waste acceptance criteria (WAC) for disposal in the Environmental Management Waste Management Facility (EMWMF) on the Oak Ridge Reservation and the sections that exceed the EMWMF WAC and will require disposal at the Nevada National Security Site (NNSS). Similar deterioration and technical issues are anticipated for the K-27 Building, which has a similar history to the K-25 Building. Therefore, estimates for the cleanup of K-27 have increased accordingly.

The pictures on the following page provide examples of the K-25 Building's deteriorating condition.

Withdrawal Alley Post Shoring

K-311-1 Subsidence – AB Booster Station
(Note: Equipment Out Of Plumb)

21

The Department is utilizing its risk management system to manage, mitigate, avoid, or eliminate the project risks as well as control potential project impacts. The risk management system includes provisions for identifying, managing, and tracking the risk elements. Key challenges and uncertainties specific to ETTP include the following:

- The Department and its predecessors processed recycled fuel through the GDPs, which spread activation and fission products throughout the enrichment cascades. Isotopes, such as technetium which is highly soluble in water and a mobile radionuclide, provide unique challenges in waste management, which the Department will address as appropriate.
- When the Department shut down the K-25 and K-27 buildings, the uranium was not completely flushed from the system. Therefore, the hold up of uranium in piping and process equipment is a risk that must be addressed prior to general D&D.
- Although the Department has learned a great deal about the extent and magnitude of contamination at ETTP, additional characterization is ongoing and will continue in the future. The Department and its contractors have developed DQOs in partnership with the regulators to define and scope sampling campaigns necessary to compliantly and efficiently disposition waste.
- Much of the ETTP process equipment and enabling technology remains classified from a security perspective. The availability of trained and qualified security cleared workers is a challenge to ETTP.

The following key assumptions have been made in determining the cost to complete the cleanup of ETTP:

- The site will be to meet industrial land use criteria;
- Costs for demolition of all buildings are estimated assuming they will be demolished;
- Waste generated during cleanup that meets the waste acceptance criteria will be disposed of in the Oak Ridge on-site disposal cell;
- Some building slabs and infrastructure not contaminated above remediation levels may be left in place;
- Sediments and soil to 10 ft deep at some locations will require removal;
- Some in-situ groundwater treatment may be required to protect potential receptors; and
- Long-term institutional controls will be required to protect against residential use and control access to deeper soil contamination and contaminated groundwater.

ACCOMPLISHMENTS AT ETTP

Decontamination and Decommissioning

- K-25 Building demolition is in process. The demolition of the half mile long West Wing of the K-25 building was completed in January 2010 with a total of 9,464 shipments of demolition debris, 871 compressors, and 873 converters shipped to the Environmental Management Waste Management Facility (EMWMF) from March 2009 through January 2010. The West Wing's footprint covered approximately 20 acres. The remainder of the West Wing debris will be disposed during 2010. The photographs below provide an aerial look at the accomplishments on this facility.

- Hazardous materials abatement continues at the K-27 Building.
- Approximately 100 auxiliary facilities have been demolished and have had their waste disposed of since the previous report for a cumulative total of 350 facilities.

- Historical success includes completing process equipment removal at three of the large process buildings: K-29, K-31, and K-33 in 2005. K-29 was the first process building to be demolished in 2006. K-31 and K-33 underwent D&D but have not yet been demolished. The equipment removal and decontamination at these three buildings included dismantling, removing, and disposing of all major components including 1,536 converters, 1,534 compressors, and 460 miles of piping. Over 13,100 waste shipments totaling approximately 320 million pounds were sent to either the Nevada National Security Site or Envirocare of Utah, Inc. (now Energy Solutions) disposal facilities.

Remedial Actions

- Through ongoing efforts, approximately 1,400 acres of the 2,200 acres requiring CERCLA cleanup at ETTP have been addressed as of the end of FY 2009.
- Over 32,000 cubic meters of contaminated soil and material was excavated and disposed from the K-1070-B Burial Ground since the previous report.

Historical success includes:

- Excavation of 26 trenches and 62 circular auger pits at the K 1070-A burial ground was completed in 2003, which resulted in disposal of over 17,480 cubic meters of waste at the Oak Ridge on-site disposal facility.
- Remediation was completed in 2001 at the K 1070 C/D G-Pit. The pit was considered to be a primary source of organic contamination in area soils and groundwater. The G-Pit excavation resulted in approximately 175 cubic meters of contaminated soil that was treated by thermal desorption and disposed in the Oak Ridge Industrial Landfill located at the Y-12 Plant.
- More than 15,000 tons of contaminated soils and debris have been removed from Blair Quarry, a former waste disposal site adjacent to ETTP, with site restoration completed in 2005. The project served as a pilot for the strategy that was developed and approved for the characterization and verification of cleanup of the Zone 1 and Zone 2 areas at ETTP.
- More than 48,100 tons of scrap metal have been removed from the K-1064 and K-770 Scrap Yards and disposed in the EMWMF through 2007.

VI. Portsmouth Gaseous Diffusion Plant

History

The 3,778-acre Portsmouth site is located in south-central Ohio in rural Pike County, approximately 22 miles north of Portsmouth, Ohio. Construction of the site began in late 1952. The mission of the site was to increase the national production of enriched uranium and maintain the nation's superiority in the development and use of nuclear energy. The plant enriched uranium for commercial reactor fuel and military applications.

In the mid-1980s, the facilities and equipment required for the next generation of enrichment facilities technology, the Gas Centrifuge Enrichment Plant (GCEP), were constructed and installed at Portsmouth. However, the project was terminated in 1985 before going into full production because of a significant reduction in the worldwide demand for enriched material.

From 1991 until production ceased in 2001, the site produced only low-enriched uranium for commercial power plants. In 1993, uranium enrichment operations were turned over to USEC in accordance with the EPAct. USEC operated the plants to enrich uranium as a government corporation and leased

---

CLEANUP PROGRESS AT PORTSMOUTH

- Demolished 21 inactive facilities, four since 2007
- Closed all five on-site landfills
- Completed uranium deposit removals in two of three process buildings by 2009
- Completed removal of 5.7 million pounds of lubricating oils and transformer oils (totaling 576,000 gallons) from the gaseous diffusion plant by 2009
- Removed 1.5 million cubic feet of scrap metal (8,400 tons) and legacy waste (over 50,000 containers)
  Removed 187,000 drums of lithium hydroxide monohydrate from site - material sold to private vendors
- Completed site-wide investigative studies and implemented treatment at all 5 groundwater plumes

---

the facilities from the Department. The regulation of nuclear safety during enrichment operations was transferred to the NRC in March 1997 and USEC completed the privatization process in July 1998.

In August 2000, USEC announced its intention to terminate enrichment operations at the Portsmouth GDP, and ceased these activities at the site in May 2001. At that time, the Department contracted with USEC to establish a Cold Standby (CSB) Program to maintain enrichment restart capability at the facility as a strategic hedge against disruption in the nation's supply of enriched uranium. The Department later re-evaluated the uranium market and terminated the CSB program at the end of FY 2005. Since that time, a Cold Shutdown Program (CSD) has been instituted to perform deactivation activities in preparation of planned D&D activities.

In December 2002, USEC announced that the former centrifuge buildings at the Portsmouth site would be used for a lead cascade centrifuge demonstration plant. In January 2004, USEC announced that the Portsmouth site had been selected for a new advanced centrifuge uranium enrichment commercial plant (the American Centrifuge Plant) that would use former centrifuge facilities located at the site. In February 2009, USEC announced it would moderate the growth in spending on the American Centrifuge Plant until it gained more clarity on an internal decision that is needed to finance the construction of the new plant.

Regulatory Basis

Remedial actions at Portsmouth are currently subject to a USEPA Administrative Consent Order issued on September 29, 1989 (amended in 1994 and 1997), a Consent Decree with the State of Ohio, issued on September 1, 1989, and a Director's Final Findings and Orders agreement with the State of Ohio, effective on April 13, 2010. The Administrative Consent Order and Consent Decree address the investigation and cleanup of releases to environmental media of hazardous wastes pursuant to sections 2002 (a)(1) and 3008 (h) of RCRA and state hazardous waste laws, and hazardous substances that are not hazardous wastes pursuant to section 104 and 106(a) of CERCLA. The agreements define roles and responsibilities and require the Department to investigate Portsmouth for potential environmental impacts of past operations which resulted in releases or spills of hazardous material and provide groundwater and soil remediation plans as required. The investigation of the groundwater and soil has been conducted in a phased approach by dividing the site into four groundwater quadrants (a watershed approach) based primarily on the direction of groundwater and surface water flow. The April 2010 Director's Final Findings and Orders agreement addresses the investigation and cleanup of the GDP buildings and facilities (non-environmental media), as well as evaluation of site-wide waste disposition and establishes the agreed upon regulatory framework for conducting those activities under CERCLA. Other agreements and permits have also been negotiated for Portsmouth to ensure compliance with state and federal laws and regulations (e.g., RCRA, TSCA, CWA, etc.) In addition, the Department is also complying with the National Environmental Policy Act, as appropriate.

The Department and regulators are in agreement that public involvement is a vital component of a successful cleanup project. Therefore, the cleanup approach will include provisions for substantive involvement of the public and other stakeholders. Accordingly, in August 2008, the Portsmouth Site Specific Advisory Board (SSAB) was chartered to provide advice to the Department regarding environmental cleanup activities at the Portsmouth site. The primary mission of the Board is to provide informed recommendations on major issues regarding environmental restoration, waste management and related cleanup activities. In addition, there is public involvement as part of the NEPA process.

Cleanup Plan, Cost, and Schedule

The Portsmouth D&D Project includes demolition and disposal of the GDP process equipment, process buildings, and other facilities auxiliary to the gaseous diffusion process. The project also includes remediation of contaminated soils and groundwater associated with gaseous diffusion uranium enrichment operations. The objective of the project is to eliminate the potential for future contaminant releases from the GDP site in a manner that protects the site workers, off-site human health, and the environment. The end-state vision is a cleanup for industrial land use of the majority of the site, but the Department, public, and regulators must work together to establish the specific completion criteria. A total of 134 facilities at the GDP that include approximately 10,600,000 square feet of floor space will be addressed by the project.

Most of the Portsmouth GDP facilities are under lease to USEC and must be returned to the Department for D&D. Although the three process buildings were returned on September 30, 2010, many facilities – including feed and withdrawal buildings, and other support facilities – are planned to be returned in late 2010. As to the remaining facilities, USEC may elect to continue leasing certain GDP facilities that are needed to support the American Centrifuge Plant (ACP).

Overall, the Portsmouth D&D Project is organized for descriptive purposes into nine distinct phases: (1) engineering and preplanning; (2) initial facilities S&M; (3) facility characterization; (4) hazardous materials abatement; (5) process equipment removal; (6) facility demolition; (7) characterization and remediation of soil and water from the deferred units; (8) disposition of process equipment, demolition debris, and contaminated soils; and (9) project closeout and transition to a long-term stewardship organization. A projection of estimated waste quantities is provided on the next page.

**ESTIMATED PORTSMOUTH GASEOUS DIFFUSION PLANT
D&D WASTE QUANTITIES AND MAGNITUDE COMPARISIONS**

Preliminary D&D Waste Sources	Estimated Waste Quantity	Size Comparisons
D&D Waste	1,700,000 m³	54,000 railroad cars making a 500-mile long train stretching from Portsmouth to Washington, D.C.
Remedial Action Waste	450,000 m³ of soil	13,000 railroad cars making a 123-mile long train
Total Number of Buildings / (Sq. Ft.)	134 / 10,600,000 ft²	Twice the space of the Pentagon
Electrical Wiring	4,600 miles	Would stretch from Portsmouth to the Pacific Ocean and back
Process Piping	600 miles	Would stretch from Portsmouth to the Atlantic Ocean

While the phases of the D&D Project are generally considered sequential for a particular facility, activities under each phase may be ongoing simultaneously in some facilities because of the massive size of some buildings. D&D of the Portsmouth GDP is scheduled to be conducted from FY 2009 through 2044. A mission need for D&D of the Portsmouth GDP was approved in 2005 and an alternative selection and cost range was approved in August 2007. The alternative selection and cost range approval is referred to as "Critical Decision 1." This approval was granted by the Department in 2007 to initiate the Portsmouth D&D project. Award of a D&D contract to initiate major D&D work was made in August FY 2010.

To plan and manage the D&D contractor's work scope in the project execution phase, the Department is developing a Performance Baseline. Concurrent with established Performance Baseline development, the Department is also preparing project execution phase Start of Process Equipment D&D. Both the baseline and process equipment D&D shall undergo a review and approval process prior to implementation.

As part of the planning process, the Department arranged with the USACE to prepare independent, comprehensive cost estimates to complete D&D of the GDP facilities at both Portsmouth and Paducah. These cost estimates were developed for planning and budgeting purposes and are conceptual in nature. The original estimated cost range for the Portsmouth project was approximately $5 billion to $12 billion (year of expenditure). Each estimate for the project cost baseline includes project management, preparation of regulatory and other planning documents, facility characterization, utility reconfiguration, waste management, D&D/disposal of all GDP facilities, characterization and remediation of the deferred units, management and oversight, and project closeout. Long-term stewardship (LTS) obligations are not included in the estimates.

The estimate used in this Triennial Report is predominantly based on the USACE of Engineers estimates. For FY 2010 and forward, the estimate to complete the cleanup of the Portsmouth GDP is $7.5 billion in year of expenditure dollars. This total consists of approximately $7.3 billion from the Fund, $111 million from remaining ARRA funds, and $100 million from uranium transfer funding during FY2010[1]. The $7.5 billion in year of expenditure dollars includes a range of $5.3 billion to $11.3 billion which reflects the +50% to -30% standard range for conceptual cost estimates for construction projects. It is important to note that this standard range has not always bounded the full cost range for nuclear facility decommissioning projects. The early D&D work has been focused on support facilities, and the D&D of the process buildings is planned to begin in 2011 under the new Portsmouth D&D contract. The new contract includes requirements to update and refine the Portsmouth D&D cost estimates, and to identify and mitigate uncertainties in the cost estimates.

Some early deactivation and pre-decommissioning of Balance of Plant facilities have been accomplished as these facilities are returned from USEC. Through the funds provided by ARRA, Portsmouth has been able to accelerate cleanup and initiate demolition of 12 facilities. Likewise, projects for remediation of a groundwater contamination plume and repackaging and dispositioning excess uranium are also under way. A separate Department of Energy project is also underway at Portsmouth to convert stored depleted uranium from its current hexafluoride form to a more stable oxide form for storage or disposal. The depleted uranium conversion project is separately funded and is not receiving D&D Funds. A projected schedule for the overall Portsmouth cleanup projects is provided on the next page.

---

[1] The $100 million in uranium transfer funding is comprised of transfers made in FY2010 to USEC for accelerated cleanup work at the Portsmouth site under the Cold Shutdown contract. One additional transfer to USEC is scheduled for the last quarter of calendar year 2010, for a total of $130 million.

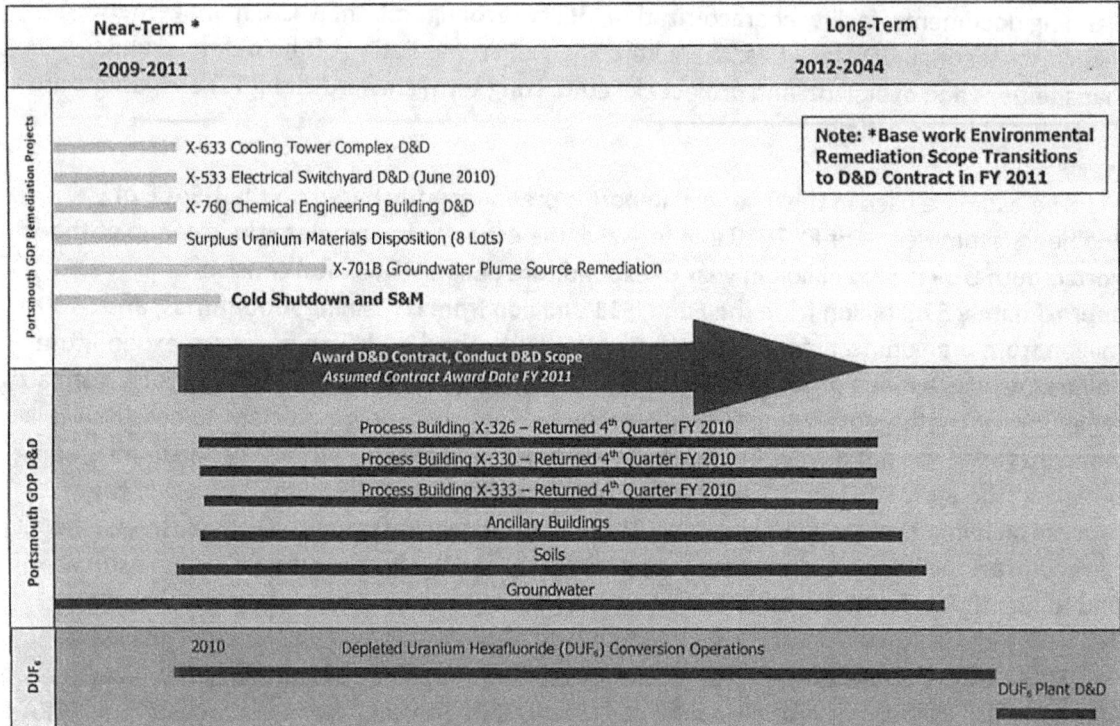

Portsmouth Cleanup Project Summary Schedule

Recent Developments at Portsmouth

The Portsmouth D&D project is in the midst of a critical development phase. The estimates and assumptions in this report are largely consistent with a 2006 conceptual cost estimate, but the understanding of the project and the long-term cleanup vision for the Portsmouth site has been maturing as the Department has been interacting with the public, the regulators, and the newly formed site specific advisory board. In addition, the Department has been able to identify and develop mitigation measures for technical issues that have been encountered at the ETTP D&D project. The Department awarded the Portsmouth D&D Project contract in August 2010. The new contractor will be working with the Department to establish a validated performance measurement baseline for the D&D project during 2011. This baseline will include updated information and assumptions that were not accounted for in the 2006 conceptual cost estimate. Some of the new factors will increase the cost estimates, and others will lower the cost estimates. There are several examples of potential cost estimate refinements, which include:

- The Department is chemically treating deposits of uranium and technetium in the Portsmouth system based upon lessons learned from Oak Ridge. These treatments are increasing near term costs, but will mitigate future project risks and are anticipated to reduce future D&D costs.

- The conceptual cost estimate assumed that almost 100 percent of the D&D waste would be disposed of in an on-site waste disposal facility as a "bounding" assumption. The potential for on-site disposal of some of the project waste is under discussion, but it has not been approved. Therefore, for the early years of the project, all waste disposal will need to be transported to off-site facilities.
- There has been interest in asset recovery from the local community as an alternative to disposal of all D&D materials as waste, but this was not included in the conceptual cost estimate.
- The conceptual cost estimate included aggressive assumptions for reductions of site infrastructure costs during D&D work, but the pace and extent of these cost reductions remains uncertain.

Challenges and Uncertainties

A variety of challenges and uncertainties exist for D&D efforts at Portsmouth. Some of the major uncertainties associated with the Portsmouth cleanup include the following:

- Disposition of the waste from the project-waste from Portsmouth is currently disposed of at off-site locations. The cleanup planning and stakeholder/regulatory review process will include a formal evaluation of waste disposal options from the project including an option to dispose of some of the waste at an on-site facility.
- The uncertain status of the USEC American Centrifuge Plant has contributed to a lack of clarity on USEC plans and schedules to return other Portsmouth facilities (some of the other support facilities at Portsmouth may be needed to support centrifuge operations).
- Level of contamination under the operating facilities and the necessary scope of remediation - currently it is not known what level of contamination, if any, is present under the GDP operating facilities. Schedules and estimates may be impacted if significant contamination is found during the D&D of these buildings.
- Asset recovery – there is uncertainty regarding the potential treatment of materials prior to, or as an alternative to, disposal. The Department is evaluating the potential melting of metal into ingots to realize technical, socioeconomic, and environmental benefits.

Lessons Learned

As part of the D&D planning process, the Department has included relevant recommendations from past reports (e.g., National Academy of Sciences study for reducing D&D costs for the three GDPs) and other applicable lessons learned in an effort to maximize the efficiency and cost effectiveness of remedial actions and D&D activities at the site. Specific examples of steps taken in response to past recommendations to maximize the efficiency and cost effectiveness of remedial actions and D&D activities at the site include the following:

- Stakeholder involvement to assure meaningful participation in the remedial action and D&D process;
- Coordination with state and federal agencies to ensure consistency in the regulatory approach and implementation of D&D;
- Completion of a plan, schedule, and cost estimate for remedial actions and D&D activities at Portsmouth;
- Incorporate both GAO and National Academy of Public Administration recommendations that large projects be divided into Capital Asset and Operations Activities so that progress can be more readily assessed in each category and during each phase; and
- Ongoing prioritization of cost- and risk-reduction factors in remedial action and D&D planning.

In keeping with the agency-wide emphasis on sound project management practices, the Portsmouth D&D Project will integrate lessons learned from other D&D sites. In addition, the proposed project schedule has been planned to allow the utilization of many members of the existing trained and cleared work force.

Accomplishments of the Portsmouth Cleanup to Date

Since the early 1990s, the Department has been conducting a comprehensive environmental cleanup program at the Portsmouth GDP. Significant actions initiated and completed to date are discussed in the following sections.

Decontamination and Decommissioning

Several major accomplishments achieved under the D&D efforts at the Portsmouth site include the following:

- Removed hazardous and flammable lube and pyranol and transformer oils from the X-333 and X-330 process buildings in 2009;
- Demolished 21 inactive facilities, four since 2007;
- Completed removal in 2008 of 438 converter shells stored outside since the 1970s. The shells were sheared, size-reduced and shipped to the Nevada National Security Site;
- Completed the Tc-99 decontamination project recovering 15,242 tons of previously unusable uranium in 2010, including 173 tons of additional uranium beyond the original scope;
- Removed deposits with criticality concerns (all 17 greater-than-safe-mass uranium deposits removed) in the X-333 and X-330 process buildings, in 2008 and 2009, respectively;
- Initiated cleanout of Department of Energy Material Storage Areas (DMSAs) 11 and 12 in 2008 (over 266.5 tons of material have been shipped to date);

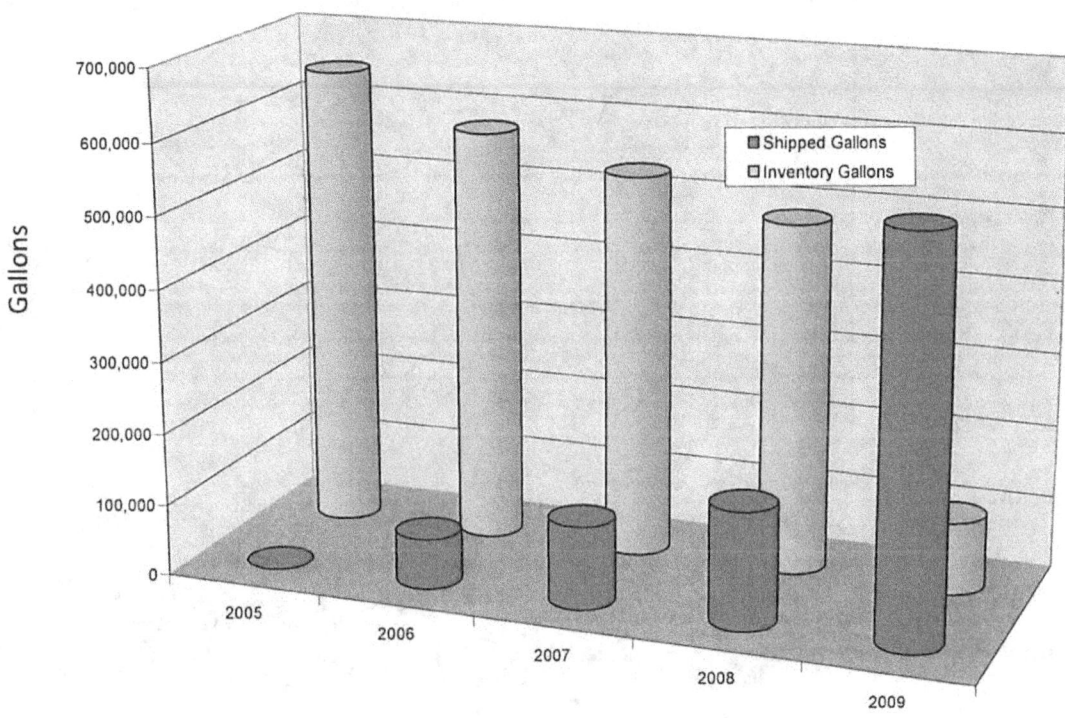

PORTSMOUTH LEGACY OIL DISPOSITION

- Worked with the community to form the Portsmouth Site Specific Advisory Board (SSAB) in 2008,
  - The Board has created two committees, the D&D Committee and the Future Land Use Committee;
    - Established web site to keep public informed on SSAB activities; and
    - Established permanent meeting location and regularly-scheduled meetings;

Remedial Actions

- Completed off-site disposition of stored legacy waste in 2007 (the site had accumulated 50,000 containers of legacy waste as of 2000).
  - Legacy waste was shipped to Energy Solutions in Utah, the Nevada National Security Site or the TSCA incinerator in Tennessee (waste is now removed as it is generated);
- Capped and closed all five site landfills (from former operations);
- Installed numerous groundwater controls and treatment facilities including wells, barrier walls, pump and treat facilities, and phytoremediation plots;
- Continued operation of four groundwater treatment facilities treating 31 million gallons annually;

Among the facilities demolished at Portsmouth was the X-633-2C Cooling Tower (top) and the Pump House (bottom) in the X-633

- Completed sitewide investigative studies and implemented treatment at all five groundwater plumes;
- Installed additional extraction wells to "pull" TCE contaminated groundwater back to the Department's property at southern plant boundary;
- Initiated real time waste shipments in FY 2008 that totaled 11,687m^3 through FY 2009; and
- Met 100% of the Site Treatment Plan milestones since the plan was approved in 1995.

VII. Paducah Gaseous Diffusion Plant

HISTORY

The Paducah Gaseous Diffusion Plant (PGDP) is located in western McCracken County, Kentucky, approximately three miles south of the Ohio River and approximately 10 miles west of the city of Paducah. The Department-owned property encompasses 3,556 acres.

The Department leases 1,986 acres outside the fenced security area to the Commonwealth of Kentucky as part of the West Kentucky Wildlife Management Area.

The site originally was known as the Kentucky Ordnance Works (KOW), a World War II munitions plant. In October 1950, the Atomic Energy Commission picked the KOW site for the second of three planned uranium enrichment plants. Operations began in 1952 and continue to produce low-assay enriched uranium for nuclear reactor fuel. The EPAct transferred responsibility for uranium enrichment at Paducah to USEC. Since the shutdown of enrichment operations at the Portsmouth GDP, Paducah has been the sole domestic producer of enriched uranium hexafluoride (UF6).

While USEC operates the enrichment facilities, the Department maintains ownership and acts as the site "landlord." USEC is responsible for the operation and maintenance of all primary process facilities and auxiliary facilities at the site. Although USEC is currently operating the PGDP, the company has announced its intent to transfer production operations to the new ACP when it becomes operational on the Department's Portsmouth reservation. USEC is expected to eventually cease operations at the PGDP. Based on the current lease renewal, the current forecast for notification from USEC of its intent to shutdown PGDP operations is 2014, with transition of PGDP leased facilities to the Department to occur in 2016.

In addition to "landlord" activities, the Department oversees environmental restoration and waste management. Waste at this site is generated from remedial actions, former enrichment operations (i.e., legacy waste generated before USEC assumed responsibility), and D&D.

REGULATORY BASIS

PGDP is in an area of abundant surface water and groundwater resources. Bordering the east and west sides of the secure area are Little Bayou Creek and Bayou Creek, respectively. Both creeks flow north toward the Ohio River and much of their flow contains permitted effluent releases from PGDP (i.e., releases in accordance with the terms and requirements of state and federal permits). These effluents constitute the majority of normal flow in the creeks.

The major groundwater resource at PGDP is called the Regional Gravel Aquifer. This aquifer originated near the southern boundary of PGDP, underlies nearly all of the secure area of the plant, and continues north to the Ohio River, into which it drains.

Historic operations at Paducah have produced contaminated areas on-site and beyond site boundaries. Principal contaminants of concern include radionuclides, trichloroethylene (TCE), PCBs, metals, and other plant-related contaminants. Through spills and disposal operations, these contaminants have entered groundwater aquifers, formed groundwater plumes, and in some cases, migrated off-site and contaminated private drinking water wells. Off-site groundwater contamination was first discovered in residential wells in 1988. Initial investigation and implementation of response actions for the contaminated groundwater was addressed by an Administrative Consent Order issued by the USEPA in 1988. In 1991, the Commonwealth of Kentucky and EPA issued a Resource Conservation and Recovery Act (RCRA) permit for storage and treatment of hazardous wastes and a Hazardous and Solid Waste Amendments permit for corrective action of Solid Waste Management Units (SWMUs). In May 1994, the Paducah site was placed on the EPA NPL under the CERCLA. As a result, a Federal Facility Agreement (FFA) was signed by the Department, the Commonwealth of Kentucky, and EPA Region 4, which establishes the framework for remedial action and D&D activities at Paducah, institutes enforceable milestones for key remedial action and D&D activities, and coordinates site-specific cleanup requirements under CERCLA and RCRA.

CLEANUP PLAN, COST, AND SCHEDULE

Site cleanup at PGDP will be accomplished using a two-phased approach. The initial phase includes remediation of contaminated media, waste disposition, and D&D of current inactive facilities and is scheduled for completion by 2019. The second phase includes work scope associated with D&D of the GDP when operations cease. The second phase for Paducah D&D is scheduled to begin in 2017. Remedial action and D&D activities for both phases will be completed under the Paducah FFA process in cooperation with the Commonwealth of Kentucky and EPA.

The technical approach for remedial action includes a multi-phase process that incorporates mitigation of immediate risks (both on- and off- site); reduces further migration of the high concentration portion of off-site contamination; establishes a series of operable units (OUs) to be addressed by CERCLA response actions; and concludes with a Comprehensive Sitewide Operable Unit (CSOU) that consists of a sitewide baseline risk assessment to: 1) evaluate any residual risk remaining at the site following remedial action and plant D&D, and 2) determine whether additional cleanup is necessary to address any such residual risks. This risk assessment and performance of any necessary follow-on cleanup actions will serve as the basis for NPL delisting.

This approach is being implemented through an OU framework intended to maximize opportunities from regional approaches and economies of scale, reduce documentation costs, and provide a better process to evaluate cumulative effects in all media. Although D&D activities for inactive PGDP facilities are underway, the D&D OU cannot be completed until the plant is no longer operating and facilities can be shutdown and removed.

Pre-PGDP shutdown remediation for contaminated media is scheduled for completion by 2019. The scope of the pre-PGDP shutdown D&D OU currently includes 20 inactive facilities. Seventeen inactive facilities have been completed. Three are in process (one holding pond, the interior components of the C-410/420 Complex, and the C-340 Plant).

The Paducah D&D scope includes the PGDP process buildings as well as the ancillary GDP facilities and supporting utilities and infrastructure. D&D scope therefore encompasses 532 structures, including 419 buildings with nearly 8,570,526 square feet of floor space and 113 ancillary facilities. The D&D activities are projected to include the removal of all building superstructures, concrete slabs on grade, and building foundations (removal of the slab and substructures to four feet below grade).

Site cleanup activities will occur in a sequenced approach consisting of 1) pre-shutdown scope, 2) post-shutdown scope, and 3) CSOU scope. Upon plant shutdown, the FFA parties intend to commence detailed planning to further define the implementation approach, which then will be included in the appropriate annual update of the Site Management Plan to the FFA. The final CSOU evaluation will occur following plant shutdown and completion of D&D of the GDP, D&D of the Depleted Uranium Hexafluoride Conversion Plant, and completion of post-shutdown cleanup of each of the specific OUs (e.g., GDP Groundwater Sources OU, Soils and Slabs OU). Any required environmental cleanup or monitoring will be conducted consistent with the selected remedies. Upon completion of cleanup, the site would be eligible for consideration for NPL delisting.

Since the publication of the 2007 Triennial Report, various changes in the D&D basis-of-estimate for the PGDP have been incorporated into the D&D estimates. The largest change resulted from an additional $77 million in remaining ARRA funding, allowing Paducah to accelerate D&D work scope to the near-term, with projected cost savings of approximately $28 million. Other changes in the project schedule have resulted in cost changes for S&M, waste storage, and groundwater treatment operations.

As part of the planning process, the Department arranged with the United States Army Corps of Engineers to prepare an independent, comprehensive cost estimate to complete D&D of the GDP facilities at both Portsmouth and Paducah. The original cost range for the Paducah project was approximately $5.8 to $12.5 billion (year of expenditure). This cost estimate was developed for planning and budgeting purposes and is conceptual in nature. It was developed after on-site facility review and evaluations of other completed or ongoing D&D work in Oak Ridge, Fernald, Rocky Flats, etc.

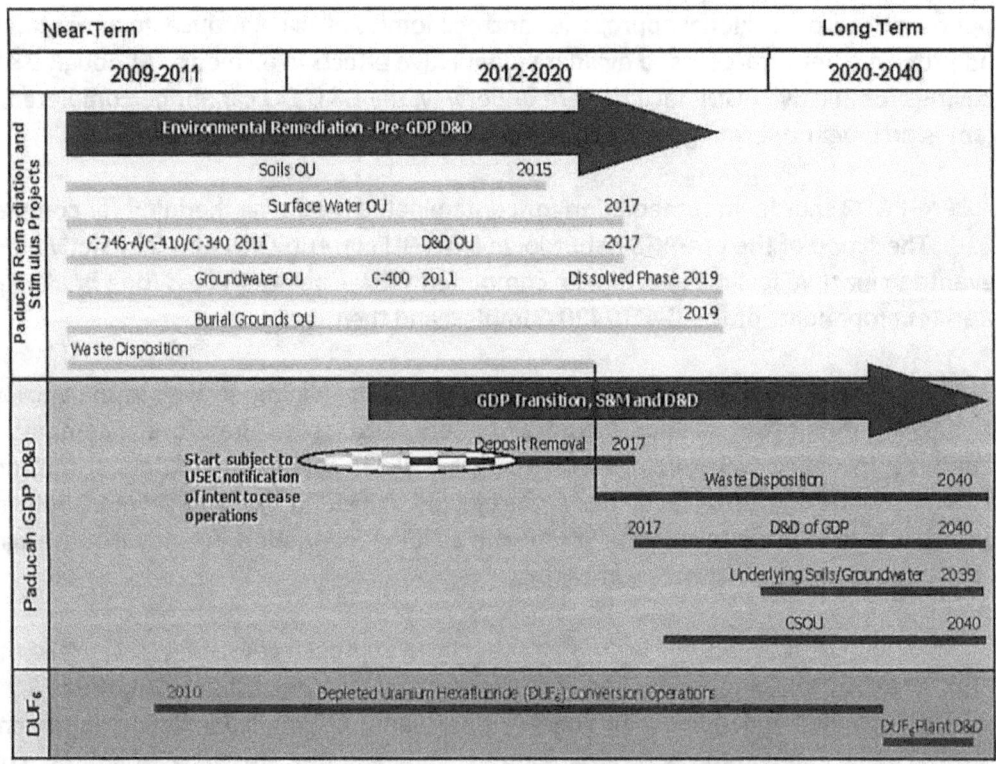

Figure 2. Paducah GDP D&D/Remedial Projects Summary

The schedule for completion of the Paducah work is Fiscal Year 2040, as shown in Figure 2 (above). Combining the conceptual estimate from the Corp of Engineers with estimates for ongoing remedial action work that is occurring prior to site D&D yields an estimate of $9.0 billion in year of expenditure dollars. This total consists of over $8.9 billion from the Fund and $77 million from remaining ARRA funding, as shown in Appendix B. The $9.0 billion includes a range of $6.3 billion to $13.5 billion which reflects a standard range of +50% to -30% for conceptual cost estimates for construction projects. It is important to note that this standard range has not always bounded the full cost range for nuclear facility decommissioning projects.

CHALLENGES AND UNCERTAINTIES

A variety of challenges remain for the D&D efforts at Paducah. Burial grounds, various contaminants in creeks and soil, off-site groundwater plumes, and on-site sources of groundwater contamination continue to pose major problems. The Department is, however, actively addressing these issues.

Several key uncertainties associated with the Paducah D&D and remedial action estimate and projected implementation plan are as follows:

- Timing for GDP transfer to the Department—it is assumed USEC will be notifying the Department of its intent to return the Paducah GDP to the Department in 2014. The time frame for this notification may change should USEC decide to extend the GDP operations.
- Scope of remedial actions for burial grounds and off-site groundwater plumes—estimates are based on volume assumptions that may change once the CERCLA decision process is completed.
- Construction of CERCLA cell—current estimate assumes on-site disposal in an approved CERCLA cell at the site. The decision process has not been completed, and the evaluation of site waste disposal options, along with public input, will affect the final decision.
- Level of contamination under the operating facilities—currently it is not known what level of contamination, if any, is present under the GDP operating facilities. Schedules and estimates may be impacted if significant contamination is found during the D&D of these buildings.

Specific examples of plans developed in response to past recommendations in an effort to maximize the efficiency and cost effectiveness of remedial actions and D&D activities at the site include the following:

- Completion of a comprehensive plan, schedule, and cost estimate for remedial actions and D&D activities at Paducah;
- Early preparation of Critical Decision (CD) Process Documents to assist baseline refinement and early GDP D&D preparation activities;
- Incorporation of both GAO and National Academy of Public Administration recommendations that large projects be divided into Cleanup Projects and Operations Activities so that progress can be more readily assessed in each category and during each phase;
- Ongoing prioritization of cost- and risk-reduction factors in remedial action and D&D planning;
- Stakeholder involvement to assure meaningful participation in the remedial action and D&D process; and
- Coordination with State and Federal agencies to ensure consistency in the regulatory

approach and implementation.

Similar to incorporation of recommendations from past reports, the Department has evaluated and implemented lessons learned from other D&D projects (e.g., East Tennessee Technology Park, Portsmouth GDP, and Savannah River). Communication and coordination with personnel from these and other Department of Energy sites undergoing D&D will continue.

ACCOMPLISHMENTS OF THE PADUCAH CLEANUP TO DATE

Paducah continues to make considerable progress toward achieving the ultimate end-state of the plant. Significant actions initiated and completed to date are discussed in the following sections.

Decontamination and Decommissioning

- Demolished two 66-year-old concrete water towers built for a WWII era munitions plant in 2009.
- Completed structure demolition and transfer for reuse of large tanks from the C-342 Ammonia Disassociator Facility in 2008.
- Completed structure demolition and disposal for the C-612 storage facility and the C-746-A West End Smelter in 2009.
- Removed asbestos-containing material from all 65 zones in the C-410 Uranium Hexafluoride Feed Plant complex by the end of FY 2009.
- Completed D&D activities at the C-603 Nitrogen Facilities in 2005, C-405 Incinerator in 2007, C-402 Lime House in 2006 and two guard posts (C-217 and C-219) in 2008.

Water Tower Demolition

- Demolished and disposed of four buildings, twelve 25,000 gallon tanks and associated piping at the hydrogen fluoride tank farm in 2005.

Remedial Actions

- Initiated Phase I system start-up processes in April 2010, after completing installation of underground and above ground treatment components for remediation of TCE contamination in soil and groundwater at the C-400 Building to recover over 50,000 gallons of TCE. This will address the largest source of off-site groundwater contamination at the site.
- Completed removal actions in 2010 for the contamination at the C-410-B Holding Pond and C-218 Firing Range.
- Excavated and restored a half-mile long contaminated on-site ditch, known as the North-South Division Ditch in 2010.

- Completed removal and disposal in 2006 of approximately 31,000 tons of scrap metal to eliminate potential direct-contact risks to plant workers and a source of surface water

Treatment System for TCE Contamination Groundwater

contamination. This was the largest collection of scrap metal at any Department of Energy facility.

- Completed installation of a sediment control basin in 2002 at Outfall 001 to control the potential migration of contamination during scrap removal.

- Applied in-situ treatment of TCE-contaminated soil at the cylinder drop test site using innovative technology (i.e., LASAGNA™ technology) to eliminate a potential source of groundwater contamination in 2002.

- Continued operations of groundwater treatment systems for both the northwest and northeast plumes installed in 1995 and 1997, respectively, to reduce contaminant migration.

- Extended municipal water lines in 1994 as a permanent source of drinking water to affected residents to eliminate exposure to contaminated groundwater.

- Constructed hard-piping in 1995 to reroute surface runoff around highly contaminated portions of the North-South Diversion Ditch to reduce potential migration of surface contamination;
- Completed by the end of FY 2009 hard-piping and installation of a retention basin and excavation of the on-site portions of the North-South Diversion Ditch, removing a direct-contact risk to plant workers and surface water contamination.

Waste Management

- Completed disposition in 2009 of legacy waste and Department of Energy Material Storage Area (DMSA) waste totaling more than 1.1 million ft^3.
- Completed characterization of all DMSAs (160) totaling over 830,000 ft^3 of material ahead of the September 30, 2009, regulatory milestone.
- Designed and built a leachate treatment system in 2008 for the C-746-U Landfill for recent expansion of disposal capacity.

VIII.   Fund Analysis

The purpose of this section is to assess whether the Fund will be sufficient to cover the required cleanup scope. This section begins by discussing the Fund's historical financial data and its current financial status. After discussing future work to be completed under the Fund, the approach for assessing the Fund's sufficiency is discussed. The section then provides results from the sufficiency analysis.

FUND RESOURCES

History and Status

Originally, the EPAct authorized annual deposits to the Fund of $330 million from Government contributions and $150 million from domestic nuclear utility assessments. The contributions would be made for 15 years (FY 1993 through FY 2007), and be adjusted annually for inflation. Detailed calculations, based upon historical production records and negotiations with the utilities, adjusted the annual utility contribution down to $148.6 million and increased the Government contribution to $331.4 million. Congress revised the Government contribution to $339.7 million annually (before inflation) beginning in FY 1999 and to $369.6 million annually (before inflation) beginning in FY 2002. Contributions in excess of current fiscal year funding requirements are invested in U.S. Treasury securities to earn interest.

The D&D Fund History and Status table on the next page shows historical inflows and outflows for the Fund from FY 1993 through FY 2009.

D&D Fund History and Status
(Dollars in Millions)

Fiscal Year	Utility Paid (A)	Government Contributions Paid (B)	Interest Income (Accrual Basis) (C)	Total Annual Receipts	Annual Costs	Cumulative Book Value of Fund Investments
1993	$147.9	$0.0	$0.1	$148.0	#N/A	$148.0
1994	172.5	197.2	11.2	380.9	$226.0	304.0
1995	160.4	133.7	23.8	317.9	352.7	266.5
1996	160.5	350.0	31.5	542.0	326.0	483.7
1997	164.9	386.6	49.1	600.6	207.5	884.2
1998	160.5	398.0	73.1	631.6	210.4	1,289.6
1999	158.8	398.1	92.5	649.4	219.0	1,715.2
2000	174.8	420.0	123.0	717.8	254.3	2,181.1
2001	180.6	419.1	141.7	741.4	309.7	2,591.4
2002	185.8	420.0	141.3	747.1	320.6	3,017.2
2003	188.9	432.7	135.5	757.1	357.2	3,447.3
2004	193.1	449.3	131.1	773.5	379.8	3,755.4
2005	197.4	459.3	144.6	801.3	546.4	4,027.4
2006	205.3	446.4	166.2	816.9	498.5	4,336.7
2007	213.2	452.0	182.8	848.0	530.2	4,673.0
2008	-	458.8	197.1	655.9	546.8	4,768.7
2009	-	463.0	183.8	646.8	595.8	4,871.4
Through 2009	$2,664.6	$6,284.2	$1,827.4	$10,776.2	$5,880.9	$4,871.4 (D)

Historical

Notes:
(A) The 15 year assessment period established by the Energy Policy Act of 1992 ended with the FY 2007 assessment. Through FY 2007, the utilities had paid all assessed amounts.

(B) The 15 year assessment period established by the Energy Policy Act ended with the FY 2007 assessment. Through FY 2007, the Government had not contributed $918.6 million of its required amounts. In addition, the Fund lost $658.9 million of interest earnings over the 15 years due to the Government not contributing its required amounts. The Government has continued making contributions beyond FY 2007 to satisfy its obligation to the Fund, as well as, making up the lost interest. The FY 2011 contribution will complete the additional Government contributions.

As discussed in the report narrative, the Government fulfilled a portion of its obligation by providing additional funding sources separate from its actual deposits into the Fund ($390M of ARRA and $100M from Uranium Transfer). The table reflects only the actual deposits, as it presents the Fund's invested balances.

(C) Interest is reported on an accrual basis and agrees with the audited financial statements. The accrual basis includes amounts in interest that were earned during the period but may not be collected. In addition, discounts and premiums are reported as adjustments to income ratably over the life of the security.

(D) The Fund's $10.8 billion in cumulative receipts through FY 2009 have been used for: 1) $5.9 billion in cleanup costs and 2) $4.9 billion to acquire U.S. Treasury securities.

For the 15-year contribution period that ended in FY 2007, the utilities made their required payments each fiscal year. Cumulative Government contributions, however, were $918.6 million less than required. Due to the delay in Government contributions, the Fund also lost potential interest earnings of approximately $658.9 million through FY 2007. Combining the contributions shortfall with lost interest yielded approximately $1.6 billion due from the Government to satisfy the Government's original obligation to the Fund. The Government has satisfied its obligation to the Fund by continuing to make deposits into the Fund after FY 2007 and by providing additional sources of funding toward the Fund's cleanup mission.

The D&D Fund History and Status table shows actual deposits into the Fund, interest earnings, costs paid from the Fund, and the Fund's invested balances. Through FY 2009, the Fund has amassed almost $10.8 billion in cumulative receipts. Approximately $5.9 billion of the receipts have been used to pay for cleanup work, with the remaining $4.9 billion invested to earn interest.

In addition to the Government's actual deposits shown in the History and Status table, the Government has also contributed to the Fund's cleanup mission by providing additional sources of funding totaling $490 million.

First, the Government designated $390 million from the FY 2009 ARRA funding as acceleration of D&D Fund activities. ARRA funding is a separate appropriation from the General Fund of the U. S. Treasury and is not available for investment in Treasury securities. Therefore, the ARRA funding is neither a contribution into the Fund's invested balances nor increased spending authority from those balances. ARRA funding essentially functions like a contribution, however, by paying for cleanup work that would otherwise have to be paid from the Fund. There is a dollar for dollar reduction in the required outlays from the Fund.

In addition to the ARRA funding, there is also uranium transfer funding that is paying for cleanup work and reducing the outlays needed from the Fund. During FY 2010, the Department transferred excess uranium to USEC in exchange for USEC performing accelerated cleanup work at the Portsmouth GDP. As with the ARRA funding, uranium transfer funding is not an actual deposit into the Fund. However, it functions like a Government contribution to the Fund's mission by paying for work that otherwise would have to be paid from the Fund. There is a dollar for dollar reduction in the required outlays from the Fund.

Investment Strategy

Consistent with the EPAct, Fund managers have predicated the investment strategy of the Fund on the expected cash outlays and receipts to the Fund. Funds not required for disbursement to the Department's contractors or to uranium/thorium licensees are invested in U.S. Treasury securities. An 18-month outlay schedule forecasts expected expenditures for Fund cleanup work. This schedule is analyzed along with the maturities of currently held investments and additional receipts (such as interest earnings and contributions) to formulate a strategy for investments maturity to meet the outlay requirements. The investment strategy intends to maximize the investment return on funds that are not required for disbursement.

As noted previously in the D&D Fund History and Status table, the Fund had a book value of $4.9 billion at the end of FY 2009. All but $23 million of the investments were in U.S. Treasury notes and bills. The $23 million was invested in U.S. Treasury overnight securities to provide the necessary liquidity for ongoing disbursements. The Department issues audited Fund financial statements each fiscal year.

FUND REQUIREMENTS

Remaining Work Scope

The Department has achieved several cleanup milestones with the $5.9 billion of costs incurred through FY 2009, but significant cleanup work remains. Cleanup estimates for future work are reflected in the following discussion.

East Tennessee Technology Park: The estimate to complete ETTP is approximately $2.1 billion in year of expenditure dollars with $2 billion from the Fund and $110 million from remaining ARRA funding ($118.2 million appropriated). Work should be completed in FY 2020. Appendix A includes a breakout of the required expenditures by fiscal year. The remaining major D&D work involves the four remaining process buildings (K-25, K-27, K-31, and K-33). The west wing of the K-25 building has been demolished, but D&D and demolition remain for the rest of the building. The K-27 building must still undergo D&D/demolition. ARRA funding is helping with preparation for D&D. ARRA funding will also be used to demolish the K-33 building. The Department is anticipating the K-31 building will be demolished and the report includes cost estimates that account for demolition of K-31. In addition to the major process buildings, approximately 150 other support facilities still require D&D. Also, remedial action work remains, especially within the main plant area where D&D of facilities must occur before remedial action. There will be continuing post-closure, long-term stewardship activities that will be paid from funding sources other than the Fund.

Portsmouth and Paducah: The schedule for D&D work at Portsmouth is FY 2009 through FY 2044. The cost estimate for this project is conceptual and for FY 2010 and forward has a most probable value of $7.5 billion in year of expenditure dollars. This total consists of approximately $7.3 billion from the Fund, $111 million from remaining ARRA funds, and $100 million from uranium transfer funding[2]. D&D work at Paducah is scheduled to begin in FY 2017 and continue through FY 2040. The cost estimate for this project is also conceptual. Combining the conceptual estimate for D&D with estimates for ongoing remedial action work that is occurring prior to site D&D yields an estimate of $9.0 billion in year of expenditure dollars for this report. This total consists of over $8.9 billion from the Fund and $77 million from remaining ARRA funding. Cost profiles for both Portsmouth and Paducah are provided in Appendix A.

Uranium/Thorium: The Fund must also continue reimbursements to licensees of active uranium and thorium processing sites for the portion of their remedial action costs attributable to Federally-related byproduct material. The Department's Office of Commercial Disposition Options oversees the uranium-thorium reimbursements. Through the FY 2009 payment, the

---

[2] The $100 million in uranium transfer funding is comprised of transfers made in FY2010 to USEC for accelerated cleanup work at the Portsmouth site under the Cold Shutdown contract. One additional transfer to USEC is scheduled for the last quarter of calendar year 2010 for a total of $130 million.

Department has paid approximately $605.3 million to the licensees ($573.4 million from the Fund and $31.9 million from ARRA). Estimates for the Triennial Report conservatively assume reimbursements will reach the legislative ceiling established by Congress with reimbursements continuing through FY 2024. Legislation enacted in 2002 (Public Law 107-222) was the last adjustment to the legislative ceiling and increased the maximum reimbursements to $715 million in FY 2002 dollars. This equates to $869 million when inflated to FY 2010 dollars. With the ceiling increasing each year for inflation, the Department could owe up to $291 million in year of expenditure dollars through FY 2024 ($253 million from the Fund and $38 million from ARRA). A future cost profile of uranium/thorium is provided in Appendix A.

FUND SUFFICIENCY

Uncertainties in Assessing Fund Sufficiency

Assessing the Fund's sufficiency involves several uncertainties. Cleanup work covered by the Fund involves large, complex projects, some of which will not be completed for approximately 30 years. Many of the cost estimates are conceptual in nature, so actual costs could fall within a sizable range of the point estimates used to assess the Fund's sufficiency.

Economic factors such as long-term inflation rates, as well as, the long-term rate of return on the Fund's investments can significantly impact Fund sufficiency. In addition, projects such as the Paducah GDP D&D have tentative start and completion dates that are several years in the future. Any delays in the assumed schedule would affect the Fund's bottom line because expensive S&M costs continue until D&D begins.

With an awareness of these uncertainties, the Department developed a "Base Case" of future assumptions to enable an analysis of Fund sufficiency. This Base Case reflects the most likely scenario for completing cleanup of the GDPs. It reflects the current programmatic assumptions about strategies, schedules, expected costs, etc. that have been discussed throughout this report. Appendix A provides a future cost profile for the Base Case. The Base Case uses forecasts for interest and inflation rates, as shown in Appendix B.

Forecasting Approach

To assess the Fund's sufficiency, the Department started with the existing Fund balance and added in projected annual inflows and outflows to get a projected annual running balance. Future cash outflows are based on projected annual spending from the Fund for FY 2010 and forward as cleanup work is performed and uranium/thorium licensees are reimbursed. Future cash inflows come from contributions and interest earnings for FY 2010 and forward. Projections of annual interest earnings are calculated by multiplying the running balance in the Fund by the forecasted interest rates on Treasury notes. Interest earnings are achieved as long as there is a positive balance in the Fund. As noted earlier, the Department calculated the cost estimates and interest earnings projections in year of expenditure dollars by using the rates presented in Appendix B.

Goals of the modeling were to determine: 1) the cost of remaining work in year of expenditure dollars; 2) the Fund's ending balance after all work is completed; and 3) if a Fund shortfall is projected, the year the Fund balance will be exhausted.

Results

Figure 3 summarizes the shortfall from the Fund sufficiency analysis. As of September 30, 2009, the Fund had incurred $5.9 billion in historical costs. When combined with estimated future costs from the Fund of $18.5 billion for the base case, a total life cycle cost of $24.4 billion in year of expenditure dollars is projected. Based on these cost projections, the Fund will have an $11.8 billion shortfall, with the Fund's balance being exhausted in 2020. Appendix C provides the annual Fund balances over the life of the cleanup work.

Results of Sufficiency Analysis					
Dollars in Billions and in Year of Expenditure					
Scenario	Historical Costs (FY 1993 - FY 2009)	Future Costs (FY 2010 & Forward)	Lifecycle Costs	Fund Shortfall	Year the Fund's Balance Becomes Negative
BASE CASE	$5.9	$18.5	$24.4	($11.8)	2020

Figure 3. Results of Sufficiency Analysis

The projected shortfall in the Fund has grown from the $10.9 billion shortfall projected in the 2007 triennial report. There are two main drivers. First, as discussed below, the Department has experienced cost growth on the cleanup of ETTP. While ARRA funding and proceeds from uranium sales have helped the Fund by paying for cleanup work, there is still a net increase in projected costs from the Fund compared to the previous report. Second, the 2007 report assumed that the Government would fulfill all its obligations to the Fund through actual deposits into the Fund. The 2010 report reflects the Government meeting about seven percent of its obligations through the other funding sources, ARRA and Uranium Transfer. These other funding sources are not actual deposits into the Fund and are not part of the invested balances on which the sufficiency analysis is performed. As stated above, there is a dollar for dollar reduction in the required outlays. The signification increase in projected costs and the slight decrease in projected deposits lead to lower running balances in the Fund, which significantly reduces projected interest earnings.

The closure of ETTP has proven to be more complex and is taking longer to complete than previously planned. Technical challenges coupled with the changes in methods of accomplishment of the demolition of the K-25 and K-27 process buildings are the primary contributors to schedule delays and cost increases. The schedule for completing ETTP is now estimated to be FY 2020, which is eight years longer than the 2012 closure date assumed in the 2007 report. The extended schedule to complete the closure of ETTP adds infrastructure and support costs that will be incurred while the cleanup work is being completed.

The K-25 building was closed in 1977. Minimal surveillance and maintenance was performed after shutdown, because there was no future mission request for the facility. The building's deteriorating 44 acre roof has led to significant, ongoing deterioration of the inside of the building. After a worker fell through a degraded floor in 2006, the K-25 D&D project was significantly slowed until the Department could re-plan the project. Revised plans utilize greater reliance on heavy equipment to demolish the building while minimizing labor inside the building for equipment removal. However, the Department has had to spend additional time and money to shore up the facility and ensure safe working conditions for the labor driven work that still has to be conducted inside the building.

Although a great deal of work has been accomplished on the K-25 Building, the deterioration of the building continues to accelerate and numerous difficulties and challenges still exist. For example, there are regulatory and sampling complexities with Tc-99 contamination and potential adjustments to the demolition approach in these Tc-99 areas to avoid spreading the contamination. Similar deterioration and technical issues are anticipated for the K-27 Building, which has a similar history to the K-25 Building. Therefore, estimates for the cleanup of K-27 have increased accordingly.

It should be noted that previous efforts to assess the adequacy of the Fund likewise yielded concerns about its sufficiency. In 1991, before the Fund was established, GAO used the Department's estimates to analyze the adequacy of several funding scenarios to support the GDP cleanup work anticipated under the proposed the EPAct (see GAO Report RCED-92-77BR "Uranium Enrichment – Analysis of Decontamination and Decommissioning Scenarios"). At that time, the Department's estimate for cleaning up the three GDPs was $19.1 billion in 1992 constant dollars (with DUF_6 disposition excluded as it is today).

To provide the needed funding, GAO concluded that the Fund would require an annual contribution of $500 million indexed for inflation to be sufficient to cover all cleanup work. GAO assumed the annual contribution would continue for the life of the cleanup work, which was predicted to last perhaps to 2040. The EPAct set an annual contribution level of $480 million indexed for inflation for 15 years. The Department was to formally assess the Fund's sufficiency at the end of the 15 years of contributions and determine if the Fund should be reauthorized. The amount to be collected over these 15 years would total $7.2 billion in 1992 dollars, which is significantly less than the $19.1 billion cost estimate that existed at that time.

Similarly, previous Triennial reports related concerns that the Fund would be insufficient to cover the full scope of work. In addition, in 2004, GAO issued audit report GAO-04-692, "Uranium Enrichment Decontamination and Decommissioning Fund is Insufficient to Cover Cleanup Costs" stating that the Fund is significantly insufficient under all the scenarios it assessed.

ADDITIONAL FUND UNCERTAINTIES

There is an inherent uncertainty associated with planning for large complex projects. As the projects progress through their life cycle, the Department's goal is to reduce this uncertainty with higher quality cost estimates, good-faith regulatory negotiations, use of best available technologies, effective contract acquisition processes, and project management planning, oversight, and controls.

Appendix A – Base Case Cost Profile

Dollars in Thousands and in Year of Expenditure

	Total	ETTP	Portsmouth	Paducah	Uranium-Thorium
Total For Site Completion	**$18,906,904**	**$ 2,112,929**	**$ 7,501,792**	**$ 9,001,535**	**$ 290,647**
Beyond D&D Fund Invested Balances:					
Remaining ARRA Funding (FYs 10-12)	$ 335,946	$ 109,664	$ 110,890	$ 77,292	$ 38,100
Uranium Transfer Funding - FY 10	$ 100,000	$ -	$ 100,000	$ -	$ -
Total From D&D Fund (Used in Sufficiency Analysis)	**$18,470,958**	**$ 2,003,265**	**$ 7,290,902**	**$ 8,924,243**	**$ 252,547**
FY 2010	$ 568,212	$ 248,765	$ 229,658	$ 89,789	$ -
FY 2011	$ 554,136	$ 230,300	$ 240,899	$ 82,937	$ -
FY 2012	$ 535,714	$ 219,300	$ 211,784	$ 88,775	$ 15,854
FY 2013	$ 544,095	$ 220,800	$ 210,797	$ 92,497	$ 20,000
FY 2014	$ 593,230	$ 224,600	$ 215,042	$ 133,588	$ 20,000
FY 2015	$ 657,253	$ 227,600	$ 219,342	$ 190,311	$ 20,000
FY 2016	$ 760,225	$ 230,800	$ 223,824	$ 285,601	$ 20,000
FY 2017	$ 799,483	$ 164,500	$ 228,566	$ 386,417	$ 20,000
FY 2018	$ 767,524	$ 116,800	$ 233,364	$ 397,359	$ 20,000
FY 2019	$ 707,758	$ 67,500	$ 238,264	$ 381,994	$ 20,000
FY 2020	$ 706,415	$ 52,300	$ 243,266	$ 390,849	$ 20,000
FY 2021	$ 625,556	$ -	$ 248,374	$ 357,182	$ 20,000
FY 2022	$ 619,615	$ -	$ 253,588	$ 346,026	$ 20,000
FY 2023	$ 569,586	$ -	$ 258,912	$ 290,674	$ 20,000
FY 2024	$ 575,470	$ -	$ 264,348	$ 294,429	$ 16,693
FY 2025	$ 570,514	$ -	$ 269,898	$ 300,616	$ -
FY 2026	$ 583,841	$ -	$ 275,565	$ 308,276	$ -
FY 2027	$ 597,032	$ -	$ 281,351	$ 315,681	$ -
FY 2028	$ 609,570	$ -	$ 287,258	$ 322,312	$ -
FY 2029	$ 626,338	$ -	$ 293,289	$ 333,050	$ -
FY 2030	$ 646,640	$ -	$ 289,487	$ 357,153	$ -
FY 2031	$ 631,611	$ -	$ 266,939	$ 364,672	$ -
FY 2032	$ 642,825	$ -	$ 298,155	$ 344,670	$ -
FY 2033	$ 655,563	$ -	$ 318,706	$ 336,857	$ -
FY 2034	$ 503,772	$ -	$ 161,329	$ 342,442	$ -
FY 2035	$ 389,435	$ -	$ 44,875	$ 344,559	$ -
FY 2036	$ 445,706	$ -	$ 96,849	$ 348,857	$ -
FY 2037	$ 472,774	$ -	$ 118,001	$ 354,773	$ -
FY 2038	$ 485,896	$ -	$ 123,666	$ 362,230	$ -
FY 2039	$ 509,174	$ -	$ 160,789	$ 348,386	$ -
FY 2040	$ 205,981	$ -	$ 174,700	$ 31,281	$ -
FY 2041	$ 112,584	$ -	$ 112,584	$ -	$ -
FY 2042	$ 113,238	$ -	$ 113,238	$ -	$ -
FY 2043	$ 50,239	$ -	$ 50,239	$ -	$ -
FY 2044	$ 33,952	$ -	$ 33,952	$ -	$ -

Note: This chart's format is intended to clarify that $18.9 billion is needed to complete the cleanup mission, but that approximately $18.5 billion will actually be outlaid from the D&D Fund. The $18.5 billion is the cost estimate used in the cost modeling that evaluates the Fund's sufficiency. The other funding sources are separate from the Fund and are neither deposits into nor withdrawals from the Fund. These other funding sources pay for work so that the Fund does not have to pay and are a dollar for dollar reduction in the required outlays from the Fund.

Appendix B -- Base Case Economic Data

Economic data used for the base case is presented in the table below.
Data for the last year of the forecast was used as the future rates for years beyond
FY 2020. Forecasts of Consumer Price Index - Urban inflation rates reflect
non-seasonally adjusted rates. Forecasts of U.S. Treasury Note interest rates reflect
10-year nominal Treasury Note interest rates. These inflation and interest rates were
adjusted to real rates by subtracting inflation from the 10-year T-Note nominal interest
rates. More specifically, real interest rates were calculated by the following equation:

Real Interest Rate = [(1+Nominal Interest Rate)/(1+Nominal Inflation Rate)] -1.

Base Case Economic Data			
	Consumer Price Index - Urban Inflation Rate	**10-Year Nominal Treasury Note Rates**	**10-Year Real Rates**
FY 2010	1.43%	3.71%	2.24%
FY 2011	2.02%	4.34%	2.28%
FY 2012	1.95%	4.85%	2.84%
FY 2013	2.00%	5.20%	3.14%
FY 2014	2.00%	5.30%	3.24%
FY 2015	2.00%	5.30%	3.24%
FY 2016	2.04%	5.30%	3.19%
FY 2017	2.12%	5.30%	3.12%
FY 2018	2.10%	5.30%	3.13%
FY 2019	2.10%	5.30%	3.13%
FY 2020	2.10%	5.30%	3.13%
Projections after 2020	2.10%	5.30%	3.13%

Appendix C -- Base Case Running Fund Balance

Dollars in Millions and in Year of Expenditure

Fiscal Year	Base Case Cost	Utility Contribution	Government Contribution	Interest Income	Fund Balance
Total	$ 18,471	$ -	$ 497	$ 1,403	$ (11,810)
2010	$ 568	$ -	$ 463	$ 163	$ 4,819
2011	$ 554	$ -	$ 34	$ 174	$ 4,472
2012	$ 536	$ -	$ -	$ 166	$ 4,102
2013	$ 544	$ -	$ -	$ 162	$ 3,720
2014	$ 593	$ -	$ -	$ 166	$ 3,293
2015	$ 657	$ -	$ -	$ 174	$ 2,810
2016	$ 760	$ -	$ -	$ 149	$ 2,198
2017	$ 799	$ -	$ -	$ 117	$ 1,515
2018	$ 768	$ -	$ -	$ 80	$ 828
2019	$ 708	$ -	$ -	$ 44	$ 164
2020	$ 706	$ -	$ -	$ 9	$ (533)
2021	$ 626	$ -	$ -	$ -	$ (1,159)
2022	$ 620	$ -	$ -	$ -	$ (1,778)
2023	$ 570	$ -	$ -	$ -	$ (2,348)
2024	$ 575	$ -	$ -	$ -	$ (2,924)
2025	$ 571	$ -	$ -	$ -	$ (3,494)
2026	$ 584	$ -	$ -	$ -	$ (4,078)
2027	$ 597	$ -	$ -	$ -	$ (4,675)
2028	$ 610	$ -	$ -	$ -	$ (5,285)
2029	$ 626	$ -	$ -	$ -	$ (5,911)
2030	$ 647	$ -	$ -	$ -	$ (6,557)
2031	$ 632	$ -	$ -	$ -	$ (7,189)
2032	$ 643	$ -	$ -	$ -	$ (7,832)
2033	$ 656	$ -	$ -	$ -	$ (8,487)
2034	$ 504	$ -	$ -	$ -	$ (8,991)
2035	$ 389	$ -	$ -	$ -	$ (9,381)
2036	$ 446	$ -	$ -	$ -	$ (9,826)
2037	$ 473	$ -	$ -	$ -	$ (10,299)
2038	$ 486	$ -	$ -	$ -	$ (10,785)
2039	$ 509	$ -	$ -	$ -	$ (11,294)
2040	$ 206	$ -	$ -	$ -	$ (11,500)
2041	$ 113	$ -	$ -	$ -	$ (11,613)
2042	$ 113	$ -	$ -	$ -	$ (11,726)
2043	$ 50	$ -	$ -	$ -	$ (11,776)
2044	$ 34	$ -	$ -	$ -	$ (11,810)

www.ingramcontent.com/pod-product-compliance
Lightning Source LLC
Chambersburg PA
CBHW081227170526
45165CB00009B/2985